Water and Wastewater Treatment

Water and Wastewater Treatment

Joseph Welker

CALLISTO REFERENCE

www.callistoreference.com

Callisto Reference,
118-35 Queens Blvd., Suite 400,
Forest Hills, NY 11375, USA

Visit us on the World Wide Web at:
www.callistoreference.com

ISBN: 978-1-64116-627-0 (Hardback)

Cataloging-in-Publication Data

Water and wastewater treatment / Joseph Welker.
    p. cm.
Includes bibliographical references and index.
ISBN 978-1-64116-627-0
1. Water--Purification. 2. Sewage--Purification. 3. Water treatment plants--Management.
4. Sewage disposal plants--Management. 5. Sanitary engineering. I. Welker, Joseph.
TD430 .W38 2022
628.162--dc23

Table of Contents

Preface

Water treatment is the process of improving water quality by removing contaminants and other undesired components, so that it becomes suitable for specific end-use. There are two types of water treatment, namely, drinking water treatment and industrial water treatment. The process used for removing contaminants from sewage and wastewater is known as wastewater treatment. The treatment aims at converting the wastewater into effluent so that it can be returned to water cycle. Water reclamation is the process of converting wastewater into water which can be reused. Wastewater treatment plant is the facility where the process takes place. There are various processes which are used for treatment of wastewater, such as phase separation, sedimentation, biochemical oxidation, etc. This book elucidates the concepts and innovative models around prospective developments with respect to water and wastewater treatment. Most of the topics introduced herein cover new techniques and the applications of water and wastewater treatment. This textbook will serve as a valuable source of reference for those interested in this field.

A detailed account of the significant topics covered in this book is provided below:

Chapter 1- Wastewater refers to the water which is contaminated by the activities of humans such as domestic, industrial, commercial and agricultural activities. It can be classified as sludge, greywater, sewage water, industrial wastewater, etc. This is an introductory chapter which will briefly introduce about all these classifications of wastewater as well as their impacts.

Chapter 2- The laboratory tests that are used to assess suitability of wastewater for disposal or re-use are termed as wastewater quality indicators. A few of such tests are biochemical oxygen demand, chemical oxygen demand, total organic carbon and oil and grease. This chapter has been carefully written to provide an easy understanding of these laboratory tests associated with wastewater quality.

Chapter 3- The process of removing contaminants from wastewater and converting them into reusable water is defined as wastewater treatment. Some of the wastewater treatment systems are grit removal, wastewater chlorination, activated sludge process, etc. The topics elaborated in this chapter will help in gaining a better perspective about these different systems of wastewater treatment.

Chapter 4- Biological wastewater treatment is the secondary treatment process which is used to remove sediments, oils, etc. which remain after primary treatment. These are divided into anaerobic and aerobic processes. This chapter closely examines the aspects associated with biological wastewater treatment to provide an extensive understanding of the subject.

Chapter 5- Industrial Wastewater Treatment is defined as the process which are used in treatment of wastewater produced by industries. It includes wastewater treatment in fertilizer industry, iron and steel industry, cement and ceramic industry, paper and pulp industry, etc. All the concepts related to industrial wastewater treatment have been carefully analyzed in this chapter.

Chapter 6- Agricultural wastewater treatment is the process of controlling pollution from surface run-off contaminated by chemicals present in fertilizers, pesticides and crop residues, etc. This chapter closely examines agricultural wastewater treatment to provide an extensive understanding of the subject.

It gives me an immense pleasure to thank our entire team for their efforts. Finally in the end, I would like to thank my family and colleagues who have been a great source of inspiration and support.

Joseph Welker

Chapter 1

Wastewater and its Types

Wastewater refers to the water which is contaminated by the activities of humans such as domestic, industrial, commercial and agricultural activities. It can be classified as sludge, greywater, sewage water, industrial wastewater, etc. This is an introductory chapter which will briefly introduce about all these classifications of wastewater as well as their impacts.

Wastewater refers to all effluent from household, commercial establishments and institutions, hospitals, industries and so on. It also includes stormwater and urban runoff, agricultural, horticultural and aquaculture effluent. Effluent refers to the sewage or liquid waste that is discharged into water bodies either from direct sources or from treatment plants. Influent refers to water, wastewater, or other liquid flowing into a reservoir, basin or treatment plant.

Wastewater includes dissolved contaminants, suspended solids and micro-organisms. Various levels of wastewater treatment separate the wastewater into sludge and a dissolved fraction containing much of the water, organic material, bacteria and salts. What is left behind is called sludge or biosolids. It is important to see sludge as one of the wastewater management issues for your community. Often sludge is ignored as a wastewater problem because community attention is solely focused on treating the remaining water to a level that reduces harm to waterways, and the nutrient cycle in particular.

Liquid waste, especially wastewater containing human wastes, will also produce an odour (from gases and aerosols). Odour is not a public health issue, but it can be a major source of nuisance and concern in a community. It will be part of your wastewater management challenge.

What is in Wastewater?

It is important to remember that wastewater is what goes down the pipe, and that the management of wastewater includes its impacts and infrastructural requirements. Your community therefore needs to think of itself as managing:

- The total impact of all wastewater in your surrounding catchment on public health and natural systems and processes.

- The physical systems (infrastructure) that channel some of these wastes for treatment and controlled 're-entry' back into the ecosystem.

If you take the total impact aspect first, there are probably four broad sources of wastewater in New Zealand, each with its own mix of substances that eventually find their way into wastewater:

- Household systems.

- Factories and industry.

- Commercial businesses/offices.

- Farms and horticulture.

Wastewater from farming tends to be dealt with separately. Increasingly, farmers are being required to set up on-site treatment systems for such things as animal effluent. Households, industries and commercial businesses can all use on-site systems, but often, depending on the size of the community, their wastewater will be combined and managed together.

This means that the wastewater in your area will be unique. An essential factor in determining the kinds of wastewater you will need to deal with will be the kinds of industries and processing businesses in your area. For example, if there is a local cheese factory, your wastewater system will have to deal with whey as a waste. If there is a metal-processing factory, your system may have to deal with water that has been used to wash down machinery.

Wastes from industry and businesses are known as tradewastes. It will be important to take account of these tradewastes when designing your system, and important to take account of initiatives being undertaken by industry to reduce the volume and toxicity of their wastes. It would be worth working with these industries in order to help them to deal with their own waste streams.

What is in Wastewater that Causes Problems?

Organic Material

The organic content of wastewater is made up of human faeces, protein, fat, vegetable and sugar material from food preparation, and soaps from cleaning. Some of this is dissolved into the water and some exists as separate particles.

Ecosystem Health Effects

Naturally occurring soil and water bacteria eat this organic waste and use it to grow rapidly. In a natural or dilute water environment where there is plenty of oxygen dissolved in the water, aerobic (oxygen-using) bacteria eat the organic material and form a slime of new bacterial cells and dissolved salt-waste products. If undiluted wastewater is left on its own, however, anaerobic (non-oxygen-using) bacteria decompose the waste organic material and release odorous gases such as hydrogen sulphide, as well as 'non-smelly' gases such as methane and carbon dioxide.

It is the amount of oxygen removed or the too-rapid growth of the bacterial slime that can cause the harm.

The important thing is to measure how much oxygen will be used by aerobic bacteria to convert the organic material to new bacteria. This is the 'biochemical oxygen demand' (BOD), and the standard measure is the amount of dissolved oxygen needed by aerobic bacteria over a five-day period at a water temperature of 20° Celsius (called the BOD_5). The BOD_5 strength of wastewater indicates its potential polluting impact if it is not treated. It is measured in parts per million (ppm), or in the metric system the number of grams of organic material per cubic metre (g/m^3). The BOD_5 of untreated wastewater is around 200–300g/m^3, while the BOD_5 for a healthy aquatic ecosystem would be less than 5g/m^3.

Relating these scientific measurements to everyday experience, a central issue is how much oxygen is left for fish to breathe after aerobic bacteria have used the oxygen to break down the organic material. If BOD_5 levels of less than 4 g/m³ occur in a stream that has naturally healthy levels of dissolved oxygen, then the stream system can deal with the amount of waste without affecting the fish. A good-quality healthy level of dissolved oxygen in water is around 8 to 10 g/m³. At a dissolved oxygen level of 5 g/m³ the fish become stressed, and at 2 g/m³ the fish will die from lack of oxygen unless they are able to move to more oxygenated waters.

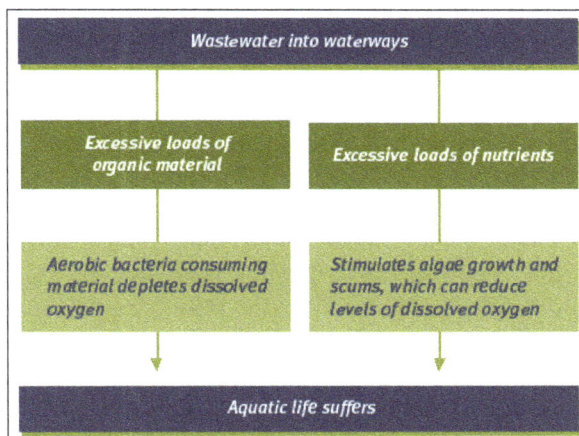

Wastewater into waterways

Excessive loads of organic material

Excessive loads of nutrients

Aerobic bacteria consuming material depletes dissolved oxygen

Stimulates algae growth and scums, which can reduce levels of dissolved oxygen

Aquatic life suffers

The effects of organic material and nutrients released into waterways.

This figure describes the effects of organic materials and nutrients released into waterways. When wastewater brings excessive loads of organic material into waterways, aerobic bacteria consuming the material deplete dissolved oxygen in the water. When wastewater brings excessive nutrients into waterways, the growth of algae and scum is stimulated, which can reduce levels of dissolved oxygen. In both cases, aquatic life suffers.

Where there is an overwhelming amount of wastewater, all the oxygen will be used up and the anaerobic bacteria will take over. The water will go septic (anaerobic) and the fish will die, as will other forms of oxygen-dependent life. This is partly why wastewater is treated to remove as much organic material as possible. But the content of even treated wastewater can be an issue for your community. Sensitive streams and estuaries are particularly vulnerable.

In effect, ecosystem services can be damaged, and these problems may be felt well before the level of pollution directly affects human health. For your area, you will need to know how much wastewater is entering—or may enter—your local stream or river, and the level of dissolved oxygen. Talking with the regional council may help with this.

Suspended Solids

The portion of organic material that does not dissolve but remains suspended in the water is known as suspended solids. The level of suspended solids in untreated wastewater is around 200 g/m³.

Ecosystem Health Effects

If effluent is discharged into streams untreated, any solids it contains will tend to settle in quiet spots. Oxygen levels will soon be depleted in the area of the contamination, causing it to decompose anaerobically. If there are high concentrations of this contamination the water in the stream will go septic because the oxygen will be used up. This will not only smother the fish, but will also kill off the life at the bottom of the stream, creating dead zones.

Dissolved Salts

The most significant salts in wastewater are nitrates and phosphates. These occur naturally to some extent. Nitrate also derives from the breakdown of organic nitrogen in protein waste matter, and the oxidation of the ammonia in urine. Phosphates are present in detergents used in washing and laundering, and are also produced by organic breakdown. The total nitrate in wastewater is around 40 g/m^3, and phosphate is around 15 g/m^3.

Ecosystem Health Effects

Nitrates and phosphates are essential elements for growth. When nitrates and phosphates are discharged into natural waters they fertilise the growth of microscopic algae and water 'weeds', which can lead to green algal suspensions and weed mats. This overgrowth results in their death and decay, and means further consumption of dissolved oxygen and smothering of aquatic life. The nutrients that caused the initial growth can then be released back into the water, initiating another cycle of weed and algal growth and decay.

Bacteria and Viruses

The human gut produces a huge quantity of bacteria, which are excreted as part of faeces on a daily basis. The most common and easily measured organism is E.coli (Escherichia coliform group), which is referred to by wastewater scientists and engineers as 'faecal coliform' bacteria. This is called an 'indicator' because its presence indicates the presence of faecal matter from warm-blooded animals. More extensive testing is required to tell if the source is human or not.

Special tests are needed to distinguish between the amount of pollution produced by humans and the amount produced by birds and other animals that gets into the water.

The amount of faecal coliform is measured per 100 ml of water—around half a cup. Each person excretes about 140 billion faecal coliforms a day. In untreated wastewater the faecal coliforms can be around 10 to 100 million per 100 ml. It is the presence of these faecal coliforms that the drinking-water standards and recreation standards are concerned with.

The main class of viruses are the enteric viruses, which cause gastro-enteritis; for example, calcivirus (Norwalk virus), rotavirus, enterovirus (polio and meningitis) and hepatitis. Generally viruses do not replicate in the outside world, but they may survive for a long time. Spray irrigation may shock viruses into die-off due to exposure to ultraviolet light or drying out of their surroundings. Poliovirus 3 has been found in aerosols at a wastewater treatment plant. In a marine environment some viruses have been known to survive a number of days, possibly protected in suspended solids.

Human Health Effects

Many of the faecal coliform bacteria in human waste are harmless. However, there are disease organisms—or 'pathogens'—that can cause harm. These can be bacteria such as typhoid, or viruses such as Hepatitis B. Direct contact with these pathogens or pollution of the water supply can cause infections. The Ministry of Health has national responsibility for developing drinking-water standards, which will guide your community's understanding of the risks it might face from local wastewater. Sewage can pollute shellfish-gathering areas and, if eaten, the shellfish will cause illness. Shellfish filter food by passing several litres of water an hour through their system. The food concentrates in the shellfish, which means that any pathogens will also accumulate.

Relatively high concentrations can also make an area unsafe for swimming and 'water contact recreation'. National guidelines developed by the Ministry for the Environment help local communities to classify their harbours, streams and lakes in terms of safety for swimming, fishing and shellfish gathering. Local regional councils will set standards for discharges for these areas. These standards relate to the amount of bacteria present in a certain volume of water.

Ground water can also become contaminated. Wastes can percolate through the soils into underground water or aquifers. Given that many smaller communities and farms obtain their water from bores or wells into these aquifers, this contamination can be a serious issue.

During the nineteenth century, the large quantities of sewage in the bigger towns and cities were identified as a health problem. Finding solutions to cholera epidemics from infected water supplies was a major issue. The wastewater system you now have may well be a direct heritage of these concerns.

Other Dissolved Constituents

Wastewater contains metals, chemicals and hormones from households (via food, medicines, cosmetics and cleaning products) and business processes (eg, mercury from dentistry, which can easily be removed by installing a centrifuge in dental surgeries). It can also contain halogenated hydrocarbons and aromatics, plasticisers, polyaromatic and petroleum hydrocarbons, organochlorine pesticides, PCBs and dioxins.

There are two issues: If large quantities are discharged into small, highly localised areas, such as a stream or small lake, there may be pollution problems. The other issue is the 'bio-accumulation' of these substances in various parts of the food chain. This can bring unacceptable concentrations in humans and aquatic life, which can lead to health problems.

Human Health Effects

Long-term Health Impacts of Residues in Water Supplies and Food

The issue here is one of long-term impacts of various wastewater residues on the human system. Water naturally contains such things as iron, zinc and manganese, but industrial processes can introduce higher concentrations. If the concentrations are high enough, exposure to some metals and chemicals may have an impact on how the body's system works.

The long-term impacts of these substances on human health are not always well understood. Wastewater will carry a range of substances, which can pass into the water supply or be returned to the soil in heavy concentrations. Some treatment systems will remove metals and chemicals from the wastewater, but the sludge produced as a result of this treatment will then contain a high concentration of these substances. The New Zealand Waste Strategy calls for such wastes, by 2007, to be beneficially used or appropriately treated to minimise the production of methane and leachate. Whatever use the sludge is put to, it should comply with the Biosolids Guidelines.

Endocrine Disruption

The endocrine system in the human body is a complex network of glands and hormones that regulate many of the body's functions, including growth, development and maturation, as well as the way various organs operate. The endocrine glands–including the pituitary, thyroid, adrenal, thymus, pancreas, ovaries and testes–release carefully measured amounts of hormones into the bloodstream, which act as natural chemical messengers. They travel to different parts of the body to control and adjust many life functions.

An endocrine disruptor is a synthetic chemical, which, when absorbed into the body, either mimics or blocks hormones and disrupts the body's normal functions. This disruption can happen through altering normal hormone levels, halting or stimulating the production of hormones, or changing the way hormones travel through the body. This is a new area of scientific investigation and is not yet well understood. There are concerns that, for example, the decline in fertility levels in all animals in the food chain, including humans, could be as a result of excessive discharge of these chemicals. Such investigations are now being considered in New Zealand.

The issue is relevant to wastewater issues because many of these substances will enter the food chain–either on land or in waterways–from wastewater. Of course some of the chemicals (eg, some pesticides) will also enter the ecosystem via run-off from farms and roadways. Wastewater treatment systems will remove some of these chemicals, but generally treatment processes are not currently designed to deal with this problem.

Ecosystem Health Effects

Endocrine Disruption

The issue raised for human health is also relevant to aquatic ecosystems. There is some concern that the hormone-producing systems in fish are under pressure. High levels of oestrogen released from wastewater can affect the reproductive cycles of fish.

Toxic Effects on Freshwater and Marine Life

These can have the immediate effect of killing fish, invertebrates and even plant life. This can be a serious loss in itself, but there are also flow-on effects. The dead fish or plants will be broken down, and can contribute to further depletion of oxygen in the water.

The key point to remember is that wastewater management is not just about toilet flushing, bathing, cooking and washing water. It is likely your community will have tradewastes, even if just from the local garage. Your overall catchment will have a huge variety of different wastewaters that will need to be considered. Table summarises the different components of wastewater that cause problems.

Table: The problem-causing Components of Wastewater.

Type of material in wastewater	Comment
Organic waste: • human waste • food waste • industrial and commercial wastes • animal effluent	• faeces, urine, blood. • an increasing volume of wastewater – possibly due to the advent of kitchen waste macerators. • tradewaste – processing. • farming – not usually managed via community infrastructure.
• Oils and fats	• households – usually from tipping down drains. • tradewastes – garages, manufacturing.
• Metals	• households – found in foods – via human wastes. • aggressive water supplies (outside the limits pH6-8). • tradewastes.
• Solvents	• households – tipping down drains, cleaning. • tradewastes – garages, manufacturing.
• Chemicals	• households – via human wastes. • households – via cleaners, soaps etc, washing, bathing and cooking. • tradewastes.
• Paints	• households. • tradewastes.

Other Effects of Wastewater

Soil Depletion

This is not so much an effect of something in the wastewater itself, but has more to do with how the management of nutrients in wastewater systems bypasses natural processes. It is worth discussing here because of the link with ecosystem health.

Over the last hundred years or so waste management design has favoured using water to transport wastes. It has also favoured direct disposal into rivers, lakes and the sea. The remaining sludge has tended to be landfilled. One effect has been to bypass the nutrient cycle, whereby wastes would be slowly returned to the soils to be taken up as a food source by plants. Some would enter the streams and rivers via groundwater but most would remain in the soils.

The depletion of nutrients from the soils has been raised as an issue in parallel with a wider concern with sustainable environmental management. This depletion means that if soils are to successfully support plant life (and farming), they must have nutrients returned through alternative processes. This can be costly.

In effect, bypassing the natural nutrient cycle means that many wastewater systems contribute to nutrient depletion in soils. Conversely, streams, rivers and lakes face risks from overloading with nutrients—with many of the problems mentioned earlier.

Soil Structure

Sediments, metals and salts can affect soil structure. For example, sodium ions can be found in high concentrations in wastewater. If irrigated on to land they can damage soil structure.

Wastewater Constituents

The constituents in wastewater can be divided into main categories according to Table. The contribution of constituents can vary strongly.

Wastewater Types

Wastewater from Society	Wastewater Generated Internally in Treatment Plants
Domestic wastewater	Thickener supernatant
Wastewater from institutions	Digester supernatant
Industrial wastewater	Reject water from sludge dewatering
Infiltration into sewers	Drainage water from sludge drying beds
Stormwater	Filter wash water
Leachate	Equipment cleaning water
Septic tank wastewater	

Constituents Present in Domestic Wastewater

Wastewater Constituents		
Microorganisms	Pathogenic bacteria, virus and worms eggs	Risk when bathing and eating shellfish
Biodegradable organic materials	Oxygen depletion in rivers, lakes and fjords	Fish death, odours
Other organic materials	Detergents, pesticides, fat, oil and grease, colouring, solvents, phenols, cyanide	Toxic effect, aesthetic inconveniences, bio accumulation in the food chain
Nutrients	Nitrogen, phosphorus, ammonium	Eutrophication, oxygen depletion, toxic effect
Metals	Hg, Pb, Cd, Cr, Cu, Ni	Toxic effect, bioaccumulation
Other inorganic materials	Acids, for example hydrogen sulphide, bases	Corrosion, toxic effect
Thermal effects	Hot water	Changing living conditions for flora and fauna
Odour (and taste)	Hydrogen sulphide	Aesthetic inconveniences, toxic effect
Radioactivity		Toxic effect, accumulation

BOD and COD

Organic matter is the major pollutant in wastewater. Traditionally organic matter has been measured as BOD and COD. The COD analysis is 'quick and dirty' (if mercury is used). BOD is slow and cumbersome due to the need for dilution series.

The COD analysis measures through chemical oxidation by dichromate the majority of the organic matter present in the sample. COD measurements are needed for mass balances in wastewater treatment. The COD content can be subdivided in fractions useful for consideration in relation to the design of treatment processes. Suspended and soluble COD measurement is very useful. Beware of the false COD measurement with permanganate, since this method only measures part of the organic matter, and should only be used in relation to planning of the BOD analysis.

The theoretical COD of a given substance can be calculated from an oxidation equation. For example, theoretical COD of ethanol is calculated based on the following equation:

$$C_2H_6O + 3O_2 \rightarrow 2CO_2 + 3H_2O$$

or, 46 g of ethanol requires 96 g of oxygen for full oxidation to carbon dioxide and water. The theoretical COD of ethanol is thus 96/46 = 2.09.

The BOD analysis measures the oxygen used for oxidation of part of the organic matter. BOD analysis has its origin in effluent control, and this is what it is most useful for. The standard BOD analysis takes 5 days (BOD_5), but alternatives are sometime used, BOD_1, if speed is needed and BOD_7 if convenience is the main option, as in Sweden and Norway. If measurement of (almost) all biodegradable material is required, BOD_{25} is used. It is possible to estimate the BOD values from the single measured value.

Relationship between BOD and COD Values in Urban Wastewater

BOD_1	BOD_5	BOD_7	BOD_{25}	COD
40	100	115	150	210
200	500	575	75	1,100

The term BOD refers to the standard carbonaceous BOD_5 analysis.

Figure below shows the dependency of time and temperature for the BOD analysis. It is important that the BOD test is carried out at standard conditions.

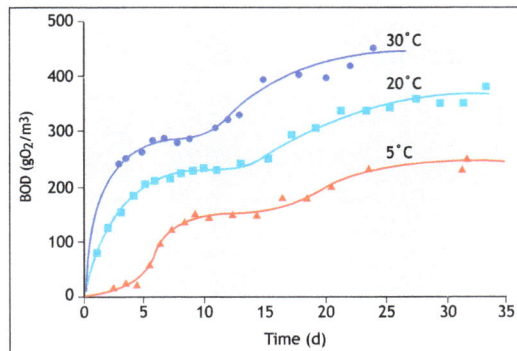

The BOD analysis result depends on both test length
and temperature. Standard is 20°C and 5 days.

Person Equivalents and Person Load

The wastewater from inhabitants is often expressed in the unit Population Equivalent (PE). PE can be expressed in water volume or BOD. The two definitions used worldwide are:

1 PE = 0.2 m³ /d

1 PE = 60 g BOD/d

These two definitions are based on fixed nonchangeable values. The actual contribution from a person living in a sewer catchment, so-called the Person Load (PL), can vary considerably.The reasons for the variation can be working place outside the catchment, socio-economic factors, lifestyle, type of household installation etc.

Variations in Person Load

Parameter	Unit	Range
COD	g/cap.d	25-200
BOD	g/cap.d	15-80
Nitrogen	g/cap.d	2-15
Phosphorus	g/cap.d	1-3
Wastewater	m3 /cap.d	0.05-0.40

Person Equivalent and Person Load are often mixed or misunderstood, so one should be careful when using them and be sure of defining clearly what they are based upon. PE and PL are both based on average contributions, and used to give an impression of the loading of wastewater treatment processes. They should not be calculated from data based on short time intervals (hours or days). The Person Load varies from country to country, as demonstrated by the yearly values given in table.

Important Components

The concentrations found in wastewater are a combination of pollutant load and the amount of water with which the pollutant is mixed. The daily or yearly polluting load may thus form a good basis for an evaluation of the composition of wastewater. The composition of municipal wastewater varies significantly from one location to another. On a given location the composition will vary with time. This is partly due to variations in the discharged amounts of substances. However, the main reasons are variations in water consumption in households and infiltration and exfiltration during transport in the sewage system.

Table: Person Load in Various Countries in kg/cap.y.

Parameter	Brazil	Egypt	India	Turkey	US	Denmark	Germany
BOD	20-25	10-15	10-15	10-15	30-35	20-25	20-25
TSS	20-25	15-25		15-25	30-35	30-35	30-35
N total	3-5	3-5		3-5	5-7	5-7	4-6
P total	0.5-1	0.4-0.6		0.4-06	0.8-1.2	0.8-1.2	0.7-1

The composition of typical domestic/municipal wastewater is shown in Table where concentrated wastewater (high) represents cases with low water consumption and/or infiltration. Diluted wastewater (low) represents high water consumption and/or infiltration. Stormwater will further dilute the wastewater as most stormwater components have lower concentrations compared to very diluted wastewater.

Table: Typical Composition of Raw Municipal Wastewater with Minor Contributions of Industrial Wastewater.

Parameter	High	Medium	Low
COD total	1,200	750	500
COD soluble	480	300	200

COD suspended	720	450	300
BOD	560	350	230
VFA (as acetate)	80	30	10
N total	100	60	30
Ammonia-N	75	45	20
P total	25	15	6
Ortho-P	15	10	4
TSS	600	400	250
VSS	480	320	200

The fractionation of nitrogen and phosphorus in wastewater has influence on the treatment options for the wastewater. Since most of the nutrients are normally soluble, they cannot be removed by settling, filtration, flotation or other means of solid-liquid separation. Table gives typical levels for these components.

In general, the distribution between soluble and suspended matter is important in relation to the characterization of wastewater.

Table: Typical Content of Nutrients in Raw Municipal Wastewater with Minor Contributions of Industrial Wastewater (in g/m^3).

Parameter	High	Medium	Low
N total	100	60	30
Ammonia N	75	45	20
Nitrate + Nitrite N	0.5	0.2	0.1
Organic N	25	10	15
Total Kjeldahl N	100	60	90
P total	25	15	6
Ortho-P	15	10	4
Organic P	10	5	2

Table: Distribution of Soluble and Suspended Material for Medium Concentrated Municipal Wastewater (in g/m^3).

Parameter	Soluble	Suspended	Total
COD	300	450	750
BOD	140	210	350
N total	50	10	60
P total	11	4	15

Since most wastewater treatment processes are based on biological degradation and conversion of the substances, the degradability of the components is important.

Table: Degradability of Medium Concentrated Municipal Wastewater (in g/m^3).

Parameter	Biodegradable	Inert	Total
COD total	570	180	750
COD soluble	270	30	300
COD particulate	300	150	450
BOD	350	0	350
N total	43	2	45

Organic N	13	2	15
P total	14.7	0.3	15

Special Components

Most components in wastewater are not the direct target for treatment, but they contribute to the toxicity of the wastewater, either in relation to the biological processes in the treatment plant or to the receiving waters. The substances which are found in the effluent might end up in a drinking water supply system in which case it is dependent on surface water extraction. The metals in wastewater can influence the possibilities for reuse of the wastewater treatment sludge to farmland. Typical values for metals in municipal wastewater are given in table.

Table: Typical Content of Metals in Municipal Wastewater with Minor Contributions of Industrial Wastewater (in mg/m^3).

Metal	High	Medium	Low
Aluminium	1,000	600	350
Cadmium	4	2	1
Chromium	40	25	10
Copper	100	70	30
Lead	80	60	25
Mercury	3	2	1
Nickel	40	25	10
Silver	10	7	3
Zinc	300	200	100

Table gives a range of hydro-chemical parameters for domestic/municipal wastewater.

Different Parameters in Municipal Wastewater

Parameter	High	Medium	Low	Unit
Absol. Viscosity	0.001	0.001	0.001	kg/m.s
Surface tension	50	55	60	Dyn/cm2
Conductivity	120	100	70	mS/m1
pH	8.0	7.5	7.0	
Alkalinity	7	4	1	Eqv/m3
Sulphide	10	0.5	0.1	gS/m3
Cyanide	0.05	0.030	0.02	g/m3
Chloride	600	400	200	gCl/m3

Wastewater may also contain specific pollutants like xenobiotics.

Table: Special Parameters in Wastewater, Xenobiotics with Toxic and Other Effects (in mg/l).

Parameter	High	Medium	Low
Phenol	0.1	0.05	0.02
Phthalates, DEHP	0.3	0.2	0.1

Nonylphenols, NPE	0.08	0.05	0.01
PAHs	2.5	1.5	0.5
Methylene chloride	0.05	0.03	0.01
LAS	10,000	6,000	3,000
Chloroform	0.01	0.05	0.01

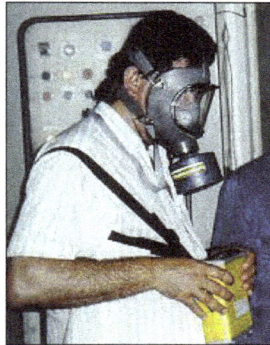

Hydrogen sulphide is often present in the influent to treatment plants, especially in case of pressurized sewers. It is very toxic and can result in casualties of personnel which do not take the necessary precautions. The picture shows measurement in the pumping station with high hydrogen sulphide concentration in the air.

Detergents in high concentrations create problems to a wastewatertreatment plant operator.

Microorganisms

Wastewater is infectious. Most historic wastewater handling was driven by the wish to remove the infectious elements outside the reach of the population in the cities. In the 19th century microorganisms were identified as the cause of diseases. The microorganisms in wastewater come mainly from human's excreta, as well as from the food industry. Table gives an idea of the concentration of microorganisms in domestic wastewater.

Table: Concentrations of Microorganisms in Wastewater (number of microorganisms per 100 ml).

Micro organisms	High	Low
E. coli	5.108	106
Coliforms	1013	1011
Cl. perfringens	5.104	103

Fecal Streptococcae	108	106
Salmonella	300	50
Campylobacter	105	5.103
Listeria	104	5.102
Staphylococus Aureus	105	5.103
Coliphages	5.105	104
Giardia	103	102
Roundworms	20	5
Enterovirus	104	103
Rotavirus	100	20

The high concentration of microorganisms may create a severe health risk when raw wastewater is discharged to receiving waters.

Surface aeration in activated sludge treatment plants creates aerosols which
contain high amount of microorganisms. This poses a health risk to treatment plant
employees and in some cases to neighbors.

Municipal Wastewater

Municipal wastewater means the mixture of domestic, process and other wastewater tributary to any given municipal sanitary sewage or treatment system.

Municipal wastewater, attributed to the widespread use of PPCPs both in the home and in health care and personal care facilities, is the primary pathway by which chemicals in prescription and over-the-counter products find their way into the aquatic environment. According to Boxall et al., regulatory environmental risk assessment approaches for PPCPs consider releases to surface waters from wastewater treatment systems, aquaculture facilities, and runofffrom fields, as well as releases to soils during biosolid and manureapplication, emissions from manufacturing sites, disposal of unused medicines to landfills, runoff of veterinary medicines from hard surfaces in farmyards, irrigation with wastewater, and the disposal of carcasses of treated animals. The release of pharmaceuticals from manufacturing facilities is heavily regulated and is not a major contributor to the environment.

Major sources of municipal wastewater are households, institutions, and commercial buildings which contain large concentrations of nutrients, especially nitrogen (N) and phosphate (P), trace elements such as iron (Fe) and Mn, dissolved salts, particularly NaCl, and in some cases, bicarbonates (HCO_3^-).

Water that falls on roofs or collects on paved areas such as driveways, roads, or footpaths is referred to as storm water. Storm water systems often run from outdoor drains down gutters and flow untreated into natural waterways(creeks, rivers, ground waters, wetlands, and oceans). Contaminants in storm water include pesticides, herbicides, oil, grease, and heavy metals such as Cd, Cr, Cu, Ni, Pb and Zn. These elements may be present in dissolved form in storm water or bound to colloidal particles.

Municipal wastewaters normally contain approximately 5% to 10% settleable suspended solids. In addition some of these solids are grease-like in nature and will separate from the denser, settleable solids. Usually cities use large settling tanks to remove both types of solids-one by the process of sedimentation and the other by floatation. Both types of solids are eventually usually wasted into the environment or treated extensively before some type of ultimate disposal is used. Both the treatment and/or the discharge into the environment are costly and damaging.

In Figure, a schematic concept of one municipal-industrial complex in which both types of solids are recovered and used by industries in the complex to make additional products is presented.

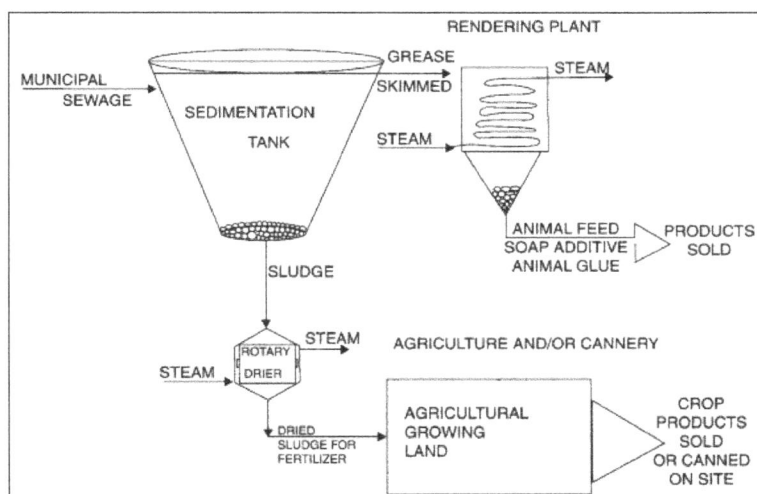

Municipal Sewage—Agriculture growing—Rendering plants.

In this complex the municipal sewage's settled solids are rotary-dried to produce a 5% to 10% cake which is then conveyed to the agricultural growing area to enhance the growth of selected fruits and/or vegetables. These food crops are then harvested and sold to outside canners or canned, if possible, onsite of the complex.

The lighter-than-water solids (grease) are skimmed from the settling tank surface and conveyed directly to the onsite renderer which then concentrates and converts these solids by enclosed heating to an edible animal food additive, natural animal glue, or a soap base for sale as products.

Transportation of the solids to distant industrial factories is avoided as is the importation of raw materials for industrial production. No municipal solids are released to the air, water, or land environments before or after costly treatments.

In addition, other advantages may exist for such a complex system. For example, the wastewater effluent from the municipal system may be reused to irrigate the agricultural growing area rather than discharging it to a nearby watercourse.

Sludge

Sludge is a semi-solid slurry and can be produced as sewage sludge from wastewater treatment processes or as a settled suspension obtained from conventional drinking water treatment and numerous other industrial processes. The term is also sometimes used as a generic term for solids separated from suspension in a liquid. Sludge can cause:

- Tube failures,

- Restricted circulation,

- Reduced system efficiency,

- Compromised boiler system reliability.

Industrial wastewater solids are also referred to as sludge, whether generated from biological or physical-chemical processes. Surface water plants also generate sludge made up of solids removed from raw water.

Sludge is a water-formed sedimentary deposit which may include all suspended solids carried by the water and trace elements in solution in the water. Sludge reduces heat efficiency and may cause scale formation in boilers and heat exchangers. Boiler sludge is a deposit that forms when suspended materials present in the boiler water settle on, or adhere to, hot boiler tubes or other surfaces.

Sludge may be formed from a combination of whatever suspended materials are in the water, including loose corrosion products, insoluble mineral precipitates and oil. A significant proportion of suspended material generated outside the boiler can find its way inside. When the amount of sludge (agglomerated suspended material) exceeds the ability of the boiler water to carry it in suspension, it settles on the boiler surfaces. A film of sludge deposited on a heat transfer surface is commonly called "baked-on sludge". It acts as a heat insulator and provides an environment promoting crevice corrosion.

Baked sludge is very difficult to remove by mechanical means, and boiler compound has no effect on it. The best method is to determine where it comes from, collect it with proper treatment and then blow it out before it cooks.

Generally corrosion and sludge build up is due to the corrosion process and the consequences of the system. The abrasive nature of sludge may lead to premature pump failure and motorized valve failure. Polymeric dispersants or sludge conditioners are added with phosphates to condition the sludge for improved removal from the boiler.

Greywater

Greywater can be defined as any domestic wastewater produced, excluding sewage. The main difference between greywater and sewage (or blackwater) is the organic loading. Sewage has a much larger organic loading compared to greywater.

Some people also categorise kitchen wastewater as blackwater because it has quite a high organic loading relative to other sources of wastewater such as bath water.

People are now waking up to the benefits of greywater re-use, and the term "Wastewater" is in many respects a misnomer. Maybe a more appropriate term for this water would be "Used Water".

Uses of Greywater

With proper treatment greywater can be put to good use. These uses include water for laundry and toilet flushing, and also irrigation of plants. Treated greywater can be used to irrigate both food and non food producing plants. The nutrients in the greywater (such as phosphorus and nitrogen) provide an excellent food source for these plants.

Benefits of Greywater Re-use

Re-using water does not diminish our quality of life, however it can provide benefits on many levels.

Two major benefits of greywater use are:

- Reducing the need for fresh water. Saving on fresh water use can significantly reduce household water bills, but also has a broader community benefit in reducing demands on public water supply.

- Reducing the amount of wastewater entering sewers or on-site treatment systems. Again, this can benefit the individual household, but also the broader community.

Treatment of Greywater Treated for Re-use

There are many ways by which to treat greywater so that it can be re-used. The various methods used must be safe from a health point of view and not harmful to the environment.

These type of greywater systems rely on plants and natural microorganisms to treat the water to a very high standard so that it can be safely re-used. The main advantage with these types of systems is that they treat the greywater naturally, and also enhance the local environment because of the attractive plants used and the fauna attracted to them.

There are other natural systems available to treat greywater. The type of system selected will depend on the specific application, and selection would be considered on a case by case basis.

Sewage Water

Sewage water is wastewater from people living in a community. It is the water released from households after use for various purposes like washing dishes, laundry, and flushing the toilet, thus the name wastewater. The used water moves from the houses through pipes installed during plumbing. The sewage water then moves into sewers, either constructed by the house owner, or into a sewer facility set up by the municipality.

Mostly, sewage water consists of grey water and black water. Grey water is the waste water from washing either from bathing, dishes or laundry. Black water is the waste water from toilets. It is characterized by debris such as paper wrappings, sanitary products, soap residues, and dirt due to the chemical composition of the various waste materials. Plus, sewage water has a foul smell.

The concern is that it due to overpopulation in urban areas without proper planning, it has resulted in Sewage pollution, which poses a threat not only to the environment but also to human health. It also affects biodiversity, aquatic life, agriculture, and is a major contributor to eutrophication and an increase in Biological Oxygen Demand (BOD).

Causes of Sewage Water

The use of Toilets as Bins

Toilets are designated as fixtures for relieving nature calls. Sadly, careless people have turned toilets into deposit banks for waste materials such as papers, sanitary products and some even go to the extent of flushing plastics. These waste materials are the causes of blockages of sewage ways in most buildings. Blockages lead to over flooding of the toilets served by that sewer.

These waste materials then result in the clogging of water ways further along sewer lines. For example, plastics such as soap wrappings clog rivers and prevent further flow due to stagnation. As a result, it harbors harmful organisms and bacteria. The blockages also lead to air pollution caused by the spread of the foul smell from the sewer. Generally, sewer treatment plants face a lot of hardship due to blockages and the foreign material present in the wastewater thereby causing sewage pollution.

Cooking Fats

Kitchen products have a lot of fats and oil. Greasy dishes are also washed in kitchen sinks. These materials are very fast at accumulating at the walls of the pipes where they form coverings that reduce the diameter of the pipe system which drain wastewater into the sewer.

Continued disposal of fats, oil and grease in sinks subsequently leads to complete blockage of the sewage system. This is even more hazardous than the blockage caused by toilets. For this blockage, no amount of cleaning can lead to the removal of the layered cover. This will warrant new installation of a pipe system, which is cost-bearing.

Also, this leads to flooding of houses as well as pollution caused by bad sewage odours. When the pressure of wastewater is high in a pipe that is clogged with fats, the obvious result is bursting of pipes which can lead to a messy situation in houses.

Overcapacity of Wastewater

Sewers are built to accommodate a certain volume of wastewater. Nevertheless, there are various reasons why sewers overflow. For starters, there are contractors who on construction of buildings end up connecting the sewage system of the new building to the existing sewer made for another residential building.

This leads to overflowing of the sewer which is hazardous to human health and can cause acute viral, bacterial and parasitic diseases like giardiasis, typhoid, gastroenteritis, and Hepatitis A.

Flooding

Flooding is also another factor that increases waste water. When there is excessive rain, as the water seeks for a pathway, it seeps into sewers and mixes with the waste water leading to more wastewater in the sewer. If the volume of the sewer is small, chances are that the sewer system will be unable to hold the increasing volume of water thereby causing sewer pollution.

Improper Handling of Wastewater

It is a practice commonly done by industries. Industries use a lot of water, and for this reason, release most of it as wastewater. As it is expected, the industries should treat the same water and bring it back into the industry machinery for reuse. However, most industries operating in areas with lax environmental policies release this raw sewage into the waterways without the least bit of treatment.

When this happens, the people living downstream suffer the most from the effects of sewage pollution. Further, there is subsequent death of aquatic life due to the release of harmful toxins that interfere with the normal activities of sea life. For instance, the release of ammonia is toxic to plants. They easily oxidize with the oxygen present in water leading to deprivation of oxygen to aquatic life.

Root Infiltration

Tree roots can be a cause of pollution of wastewater. They enter sewer lines at certain points and crack the pipes or underground sewer tanks leading to seeping of filthy waste water out of the sewers.

Industrial Wastewater

Industrial wastewater is one of the important pollution sources in the pollution of the water environment. During the last century a huge amount of industrial wastewater was discharged into rivers, lakes and coastal areas. This resulted in serious pollution problems in the water environment and caused negative effects to the eco-system and human's life.

There are many types of industrial wastewater based on different industries and contaminants; each sector produces its own particular combination of pollutants. Like the various characteristics of industrial wastewater, the treatment of industrial wastewater must be designed specifically for the particular type of effluent produced.

The amount of wastewater depends on the technical level of process in each industry sector and will be gradually reduced with the improvement of industrial technologies. The increasing rates of industrial wastewater in developing countries are thought to be much higher than those in developed countries. This fact predicts that industrial wastewater pollution, as a mean environment pollution problem, will move from developed countries to developing countries in the early 21st century.

Until the mid 18th century, water pollution was essentially limited to small, localized areas. Then came the Industrial Revolution, the development of the internal combustion engine, and the petroleum-fuelled explosion of the chemical industry. With the rapid development of various industries, a huge amount of fresh water is used as a raw material, as a means of production (process water), and for cooling purposes. Many kinds of raw material, intermediate products and wastes are brought into the water when water passes through the industrial process. So in fact the wastewater is an "essential by-product" of modern industry, and it plays a major role as a pollution sources in the pollution of water environment.

The Types of Industrial Waste Water

There are many types of industrial wastewater based on the different industries and the contaminants; each sector produces its own particular combination of pollutants.

Table: Water Pollutants by the Industrial Sector.

Sector	Pollutant
Iron and Steel	BOD, COD, oil, metals, acids, phenols, and cyanide.
Textiles and Leather	BOD, COD, solids, Chlorinated organic compounds.
Pulp and Paper	BOD, COD, mineral oils, phenols, and chromium.
Petrochemicals and Refineries	BOD, COD, mineral oils, phenols, and chromium.
Chemical	COD, organic chemicals, heavy metals, SS, and cyanide.
Non-ferrous Metals	Fluorine and SS.
Microelectronics	COD and Organic Chemicals.
Mining	SS, metals, acids and salts.

The metal-working industries discharge chromium, nickel, zinc, cadmium, lead, iron and titanium compounds, among them the electroplating industry is an important pollution distributor.

Photo processing shops produce silver, dry cleaning and car repair shops generate solvent waste, and printing plants release inks and dyes. The pulp and paper industry relies heavily on chlorine-based substances, and as a result, pulp and paper mill effluents contain chloride organics and dioxins, as well as suspended solids and organic wastes. The petrochemical industry discharges a lot of phenols and mineral oils. Also wastewater from food processing plants is high in suspended solids and organic material. Like the various characteristics of industrial wastewater, the treatment of industrial wastewater must be designed specifically for the particular type of effluent produced.

Generally, industrial wastewater can be divided into two types: inorganic industrial wastewater and organic industrial wastewater.

Inorganic Industrial Wastewater

Inorganic industrial wastewater is produced mainly in the coal and steel industry, in the nonmetallic minerals industry, and in commercial enterprises and industries for the surface processing of metals (iron picking works and electroplating plants). These wastewaters contain a large proportion of suspended matter, which can be eliminated by sedimentation, often together with chemical flocculation through the addition of iron or aluminum salts, flocculation agents and some kinds of organic polymers.

The purification of warm and dust-laden waste gases from blast furnaces, converters, cupola furnaces, refuse and sludge incineration plants, and aluminum works results in wastewater containing mineral and inorganic substances in dissolved and undissolved form.

The pre-cooling and subsequent purification of blast-furnace gases requires up to 20 m3 water per t of pig iron. On its way into the gas cooler the water absorbs fine particles of ore, iron and coke, which do not easily settle. Gases dissolve in it, especially carbon dioxide and compounds of the alkali and alkaline earth metals, if they are water-soluble or if they are dissolved out of the solid substances by gases washed out along with them.

In the separation of coal from dead rock, the normal means of transport and separation is water, which then contains large amounts of coal and rock particles and is called coalwashing water. Coal-washing water is recycled after removal of the coal and rock particles through flotation and sedimentation processes.

Other wastewater from rolling mills contain mineral oil and require additional installations, such as scum boards and skim-off apparatus, for the retention and removal of mineral oils. Residues of emulsified oil remaining in the water also need chemical flocculation.

In many cases, wastewater is produced in addition to solid substances and oils, and also contains extremely harmful solutes. These include blast-furnace gas-washing wastewater containing cyanide, wastes from the metal processing industry containing acids or alkaline solutions (mostly containing non-ferrous metals and often cyanide or chromate), wastewater from eloxal works and from the waste gas purification of aluminum works, which in both cases contain fluoride. Small and medium sized non-metallic-minerals plants and metal processing plants are so situated that they discharge their wastewater into municipal wastewater systems and have to treat or purify their effluents before discharge, in compliance with local regulations.

Organic Industrial Wastewater

Organic industrial wastewater contains organic industrial waste flow from those chemical industries and large-scale chemical works, which mainly use organic substances for chemical reactions.

The effluents contain organic substances having various origins and properties. These can only be removed by special pretreatment of the wastewater, followed by biological treatment. Most organic industrial wastewaters are produced by the following industries and plants:

- The factories manufacturing pharmaceuticals, cosmetics, organic dye-stuffs, glue, adhesives, soaps, synthetic detergents, pesticides and herbicides;

- Tanneries and leather factories;

- Textile factories;

- Cellulose and paper manufacturing plants;

- Factories of the oil refining industry;

- Brewery and fermentation factories;

- Metal processing industry.

As examples, some special types of wastewater produced by the industries mentioned above are briefly introduced as follows:

Wastewater Produced from the Pharmaceutical Industries

The quality of the wastes from the production of pharmaceuticals varies a great deal, owing to the variety of basic raw materials, working processes and waste products. It is a characteristic of the pharmaceutical industry that very many products as well as intermediate products are manufactured in the same plant. Thus different kinds of effluent with widely varying qualities flow from the different production areas.

For large chemical industries it is also usual to manufacture pharmaceutical products together with other chemical products. Some times waste substances include the extraction residues of natural and synthetic solvents, used nutrient solutions, specific poisonous substances and many other organics.

The wastewater produced by the pharmaceutical industry has a very bad quality for wastewater treatment. Usually the concentration of COD is around 5000-15000 mg/L, the concentration of BOD_5 is relative low, and the ratio of BOD_5/COD is lower than 30% which means the wastewater has a poor biodegradability. Such wastewater has bad color and high (or low) pH value, and it needs a strong pretreatment method, followed by a biological treatment process with a long reaction time.

Wastewater Produced by Tannery Plants

A tannery is one of the most water intensive plants, and its production process consists of several steps. The quality of water depends only to a slight degree on the type of hides and the mechanical and chemical methods used in tanning. In a tannery with chrome and bark tanning, the wastewater resulting from the different processes are as follows:

Process	Percentage of Wastewater Produced
Soaking and washing	22.5%
Liming	17.5%
Rinsing	5.5%
Plumping and bating	19.0%
Chrome tanning	2.0%
Bark tanning	2.0%
Washing and drumming	31.5%

In fact the wastewater flow is very uneven. The peak flow can be 250% of the hourly average flow rate.

The wastewater produced by a tannery (including preparation of the hides) has a fairly acid pH and high chloride content (up to 5 g Cl/L). It contains a high concentration of COD (about 1500 – 2500 mg/L), a high amount of settable substances (10-20 g/L) and emulsified fat, and tends to form foam. The dichromate content can reach a peak value of 2000 mg/L. So the tannery wastewater is a killer to the water environment if it is discharged without good treatment.

Wastewater Produced by Brewery Industry

Barley is the most important grain used for brewing beer, with the addition of rice, oats, rye, wheat and millet. The manufacture of beer consists of three processes, which are preparation of malt from barley, preparation of beer wort and fermentation.

A part of the wastewater produced by the brewery industry comes from the processes mentioned above, which includes the washing and rinsing water to clean the barley, all machines and filters, and especially bottles and barrels. This type of wastewater contains the high concentration of suspended solids and detergents. The other part of wastewater is produced by the fermentation process, and it has a very high concentration of COD and BOD_5 caused by soluble and insoluble organics. The composition and amount of wastewater produced by different processes are shown in Table below. The characteristic of mixed brewery wastewater shows the following composition: COD: 1500-5000 mgO_2/L; BOD_5: 1000-3000 mg/L; P_{total}: 5-30 mg/L; P_{PO4}: 2-5 mg/L and settable solids: 3-30 mg/L. The brewery wastewater is approximately three to four times more concentrated than sewage. There are no toxic contaminants in brewery wastewater, and most organic substances of the wastewater are biodegradable. So after the removal of suspended solids, usually an anaerobic biological treatment process is used to reduce the organic concentration of the wastewater, and then followed by an aerobic biological treatment process to make the quality of effluent meet the discharge standards.

Composition of Wastewater Produced by Different Processes

Type of wastewater	pH	Dry residue (mg/L)	Suspended solids (mg/L)	BOD_5 (mg/L)
Barrel cleaning	7.1	980	250	21

Bottle cleaning				
Washing solution	11.5	71700	310	870
Rinsing solution	7.2	940	95	16
Filter cloth washing				
Mash filter	6.7	1070	1846	325
Cooler sludge filter	6.7	1290	456	694
Fermentation				
Fermenting without yeast	5.3	2060	3944	3550
Fermenting with yeast	5.0	------	-----	70250
Storage without yeast	6.8	1010	164	502
Storage with yeast	5.2	------	10900	84500
Beer filter	5.9	1940	37835	2000

Impact of Wastewater

Wastewater is all around you. From the water running down your shower drain to the runoff that comes from wet roads, this is a byproduct of our modern lifestyle. Thanks to advanced wastewater treatment technology, the water you drink and shower in is filtered and treated to remove any contaminants like sewage or chemicals.

Natural Bodies of Water

Both bodies of freshwater and saltwater are polluted every day by untreated wastewater. In fact, the U.S. EPA estimates that almost 1.2 trillion gallons of sewage from household and industrial sources is dumped into the nation's water every single year, or about 3.28 billion gallons a day. And, yes, that was trillion with a "T" and billion with a "B". This not only creates an unsafe environment for marine life, it creates hazards for humans as well. The importance of wastewater treatment design and infrastructure is especially relevant with bodies of fresh water, as these materials would end up in your home if water wasn't treated properly.

Groundwater and Water Tables

Numerous parts of the world are currently suffering from water scarcity (and that includes U.S. states like California), which means clean water is of the utmost importance. When wastewater is discharged on these dry lands, it can seep into the underground water tables and well sources. Because we need to draw from these natural bodies of water for generations to come, this can render entire water supplies useless for people in multiple locations.

Natural Ecosystems

Every ecosystem relies on water in some regard. And when water is contaminated by sewage, toxic chemicals, or any number of other man-made forms of waste, those ecosystems are put at serious risk. Not only that, but surface and underground water are connected, always. Reckless disposal of waste can contaminate a far wider range of animals and environments than you may even know.

References

- Sustainable-wastewater-management-handbook-smaller-communities-part-1-0: mfe.govt.nz, Retrieved 02 January, 2019

- Municipal-wastewater, earth-and-planetary-sciences: sciencedirect.com, Retrieved 23 March, 2019

- Greywater-treatment: sustainable.com.au, Retrieved 19 May, 2019

- Sludge- 1627: corrosionpedia.com, Retrieved 03 July, 2019

- Sewage-water-treatment: conserve-energy-future.com, Retrieved 14 August, 2019

- Effects-wastewater-environment: organicawater.com, Retrieved 25 January, 2019

Chapter 2

Wastewater Quality Indicators

The laboratory tests that are used to assess suitability of wastewater for disposal or re-use are termed as wastewater quality indicators. A few of such tests are biochemical oxygen demand, chemical oxygen demand, total organic carbon and oil and grease. This chapter has been carefully written to provide an easy understanding of these laboratory tests associated with wastewater quality.

Wastewater Characteristics and Effluent Quality Parameters

Expansion of urban populations and increased coverage of domestic water supply and sewerage give rise to greater quantities of municipal wastewater. With the current emphasis on environmental health and water pollution issues, there is an increasing awareness of the need to dispose of these wastewaters safely and beneficially. Use of wastewater in agriculture could be an important consideration when its disposal is being planned in arid and semi-arid regions. However it should be realized that the quantity of wastewater available in most countries will account for only a small fraction of the total irrigation water requirements. Nevertheless, wastewater use will result in the conservation of higher quality water and its use for purposes other than irrigation. As the marginal cost of alternative supplies of good quality water will usually be higher in water-short areas, it makes good sense to incorporate agricultural reuse into water resources and land use planning.

Properly planned use of municipal wastewater alleviates surface water pollution problems and not only conserves valuable water resources but also takes advantage of the nutrients contained in sewage to grow crops. The availability of this additional water near population centres will increase the choice of crops which farmers can grow. The nitrogen and phosphorus content of sewage might reduce or eliminate the requirements for commercial fertilizers. It is advantageous to consider effluent reuse at the same time as wastewater collection, treatment and disposal are planned so that sewerage system design can be optimized in terms of effluent transport and treatment methods. The cost of transmission of effluent from inappropriately sited sewage treatment plants to distant agricultural land is usually prohibitive. Additionally, sewage treatment techniques for effluent discharge to surface waters may not always be appropriate for agricultural use of the effluent.

Many countries have included wastewater reuse as an important dimension of water resources planning. In the more arid areas of Australia and the USA wastewater is used in agriculture, releasing high quality water supplies for potable use. Some countries, for example the Hashemite Kingdom of Jordan and the Kingdom of Saudi Arabia, have a national policy to reuse all treated wastewater effluents and have already made considerable progress towards this end. In China, sewage use in agriculture has developed rapidly since 1958 and now over 1.33 million hectares are irrigated with sewage effluent. It is generally accepted that wastewater use in agriculture is justified on agronomic and economic grounds but care must be taken to minimize adverse health and environmental impacts. The purpose of this document is to provide countries with guidelines for wastewater use in agriculture which will allow the practice to be adopted with complete health and environmental security.

Characteristics of Wastewaters

Municipal wastewater is mainly comprised of water (99.9%) together with relatively small concentrations of suspended and dissolved organic and inorganic solids. Among the organic substances present in sewage are carbohydrates, lignin, fats, soaps, synthetic detergents, proteins and their decomposition products, as well as various natural and synthetic organic chemicals from the process industries. Table below shows the levels of the major constituents of strong, medium and weak domestic wastewaters. In arid and semi-arid countries, water use is often fairly low and sewage tends to be very strong.

Table: Major Constituents of Typical Domestic Wastewater.

Constituent	Concentration, mg/L		
	Strong	Medium	Weak
Total solids	1200	700	350
Dissolved solids (TDS)	850	500	250
Suspended solids	350	200	100
Nitrogen (as N)	85	40	20
Phosphorus (as P)	20	10	6
Chloride	100	50	30
Alkalinity (as CaCO3)	200	100	50
Grease	150	100	50
BOD5	300	200	100

Municipal wastewater also contains a variety of inorganic substances from domestic and industrial sources, including a number of potentially toxic elements such as arsenic, cadmium, chromium, copper, lead, mercury, zinc, etc. Even if toxic materials are not present in concentrations likely to affect humans, they might well be at phytotoxic levels, which would limit their agricultural use. However, from the point of view of health, a very important consideration in agricultural use of wastewater, the contaminants of greatest concern are the pathogenic micro and macro-organisms.

Pathogenic viruses, bacteria, protozoa and helminths may be present in raw municipal wastewater at the levels indicated in Table below and will survive in the environment for long periods, as summarized in a Table below. Pathogenic bacteria will be present in wastewater at much lower levels than the coliform group of bacteria, which are much easier to identify and enumerate (as total coliforms/100ml). Escherichia coli are the most widely adopted indicator of faecal pollution and they can also be isolated and identified fairly simply, with their numbers usually being given in the form of faecal coliforms (FC)/100 ml of wastewater.

Table: Possible Levels of Pathogens in Wastewater.

Type of Pathogen		Possible concentration per litre in Municipal Wastewater
Viruses	Enteroviruses	5000
Bacteria	Pathogenic E. coli	
	Salmonella spp.	7000
	Shigella spp.	7000
	Vibrio cholerae	1000
Protozoa	Entamoeba histolytica	4500
Helminths	Ascaris Lumbricoides	600
	Hookworms	32
	Schistosoma mansoni	1
	Taenia saginata	10
	Trichuris trichiura	120

Table: Survival of Excreted Pathogens (At 20-30°C).

Type of pathogen	Survival times in days			
	In faeces, nightsoil and sludge	In fresh water and sewage	In the soil	On crops
Viruses				
Enteroviruses	<100 (<20)	<120 (<50)	<100 (<20)	<60 (<15)*
Bacteria				
Faecal Coliforms	<90 (<50)	<60 (<30)	<70 (<20)	<30 (<15)
Salmonella spp.	<60 (<30)	<60 (<30)	<70 (<20)	<30 (<15)
Shigella spp.	<30 (<10)	<30 (<10)	-	<10 (<5)
Vibrio cholerae	<30 (<5)	<30 (<10)	<20 (<10)	< 5 (<2)
Protozoa	<30 (<15)	<30 (<15)	<20 (<10)	<10 (< 2)
Entamoeba histolytica cysts	<30 (<15)	<30 (<15)	<20 (<10)	<10 (< 2)
Helminths	Many	Many	Many	<60 (<30)
Ascaris lunbricoides eggs	Months	Months	Months	

Quality Parameters of Importance in Agricultural use of Wastewaters

Organic chemicals usually exist in municipal wastewaters at very low concentrations and ingestion over prolonged periods would be necessary to produce detrimental effects on human health. This is not likely to occur with agricultural/aquacultural use of wastewater, unless cross-connections with potable supplies occur or agricultural workers are not properly instructed, and can normally be ignored. The principal health hazards associated with the chemical constituents of wastewaters, therefore, arise from the contamination of crops or groundwaters. Hillman has drawn attention to the particular concern attached to the cumulative poisons, principally heavy metals, and carcinogens, mainly organic chemicals. World Health Organization guidelines for drinking water quality include limit values for the organic and toxic substances given in Table, based on acceptable daily intakes (ADI). These can be adopted directly for groundwater protection purposes but, in view of the possible accumulation of certain toxic elements in plants (for example, cadmium and selenium) the intake of toxic materials through eating the crops irrigated with contaminated wastewater must be carefully assessed.

Table: Organic and Inorganic Constituents of Drinking Water of Health Significance.

Organic	Inorganic
Aldrin and dieldrin	Arsenic
Benzene	Cadmium
Benzo-a-pyrene	Chromium
Carbon tetrachloride	Cyanide
Chlordane	Fluoride
Chloroform	Lead
2,4 D	Mercury
DDT	Nitrate
1,2 Dichloroethane	Selenium
1,1 Dichlorethylene	
Heptachlor and heptachlor epoxide	
Hexachlorobenzene	
Lindane	
Methoxychlor	
Pentachlorophenol	
Tetrachlorethylene	
2, 4, 6 Trichloroethylene	
Trichlorophenol	

Pathogenic organisms give rise to the greatest health concern in agricultural use of wastewaters, yet few epidemological studies have established definitive adverse health impacts attributable to the practice. Shuval et al. reported on one of the earliest evidences connecting agricultural wastewater reuse with the occurrence of disease. It would appear that in areas of the world where helminthic diseases caused by Ascaris and Trichuris spp. are endemic in the population and where raw untreated sewage is used to irrigate salad crops and/or vegetables eaten uncooked, transmission of these infections is likely to occur through the consumption of such crops. A study provides additional evidence to support this hypothesis and further evidence was also provided by Shuval et al. to show that cholera can be tranmitted through the same channel.

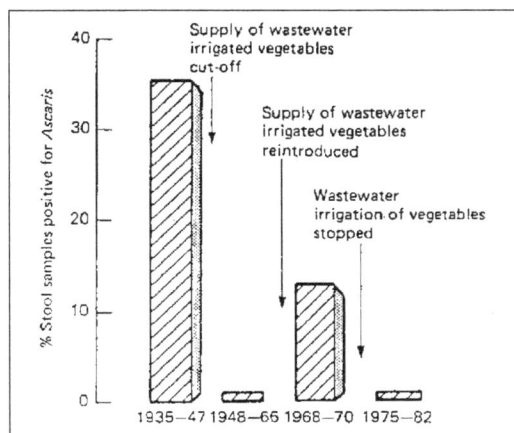

Prevalence of Ascaris-positive stool samples in West Jerusalem population during various periods, with and without supply of vegetables and salad crops irrigated with raw wastewater.

There is only limited evidence indicating that beef tapeworm (Taenia saginata) can be transmitted to the population consuming the meat of cattle grazing on wastewater irrigated fields or fed crops from such fields. However, there is strong evidence from Melbourne, Australia and from Denmark that cattle grazing on fields freshly irrigated with raw wastewater, or drinking from raw wastewater canals or ponds, can become heavily infected with the disease (Cysticerosis).

Sewage farm workers exposed to raw wastewater in areas where Ancylostoma(hookworm) and Ascaris (nematode) infections are endemic have significantly excess levels of infection with these two parasites compared with other agricultural workers in similar occupations. Furthermore, the studies indicated that the intensity of the Ascaris infections (the number of worms infesting the intestinal tract of an individual) in the sample of sewage farm workers was very much greater than in the control sample. In the case of the hookworm infections, the severity of the health effects was a function of the worm load of individuals, which was found to be related to the degree of exposure and the length of time of exposure to the hookworm larvae. Sewage farm workers are also liable to become infected with cholera if practising irrigation with raw wastewater derived from an urban area in which a cholera epidemic is in progress. Morbidity and serological studies on wastewater irrigation workers or wastewater treatment plant workers occupationally exposed to wastewater directly and to wastewater aerosols have not been able to demonstrate excess prevalence of viral diseases.

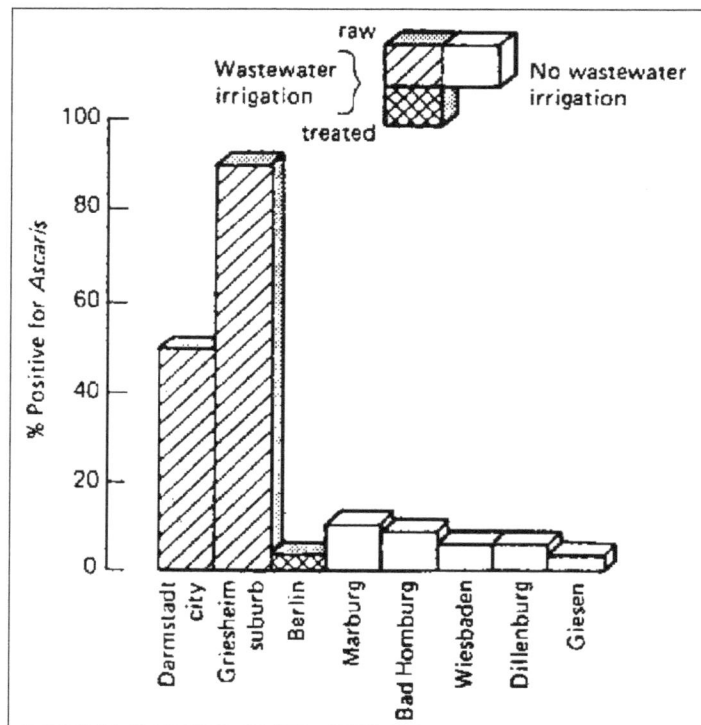

Wastewater Irrigation of Vegetables And Ascaris Prevalence In Darmstadt And Berlin, Compared With Other Cities In Germany Not Practising Wastewater Irrigation.

No strong evidence has been adduced to suggest that population groups residing near wastewater treatment plants or wastewater irrigation sites are at greater risk from pathogens in aerosolized wastewater resulting from aeration processes or sprinkler irrigation. Shuval et al. suggest that the high levels of inmunity against most viruses endemic in the community essentially block environmental transmission by wastewater irrigation.

The following microbiological parameters are particularly important from the health point of view:

Indicator Organisms

- Coliforms and Faecal Coliforms: The Coliform group of bacteria comprises mainly species of the genera Citrobacter, Enterobacter, Escherichia and Klebsiella and includes Faecal Coliforms, of which Escherichia coli is the predominant species. Several of the Coliforms are able to grow outside of the intestine, especially in hot climates, hence their enumeration is unsuitable as a parameter for monitoring wastewater reuse systems. The Faecal Coliform test may also include some non-faecal organisms which can grow at 44°C, so the E. coli count is the most satisfactory indicator parameter for wastewater use in agriculture.

- Faecal Streptococci: This group of organisms includes species mainly associated with animals (Streptococcus bovis and S. equinus), other species with a wider distribution (e.g. S. faecalis and S. faecium, which occur both in man and in other animals) as well as two biotypes (S. faecalis var liquefaciens and an a typical S. faecalis that hydrolyzes starch) which appear to be ubiquitous, occurring in both polluted and non-polluted environments. The enumeration of Faecal Streptococci in effluents is a simple routine procedure but has the following limitations: the possible presence of the non-faecal biotypes as part of the natural microflora on crops may detract from their utility in assessing the bacterial quality of wastewater irrigated crops; and the poorer survival of Faecal Streptococci at high than at low temperatures. Further studies are still warranted on the use of Faecal Streptococci as an indicator in tropical conditions and especially to compare survival with that of Salmonellae.

- Clostridium Perfringens: This bacterium is an exclusively faecal spore-forming anaerobe normally used to detect intermittent or previous pollution of water, due to the prolonged survival of its spores. Although this extended survival is usually considered to be a disadvantage for normal purposes, it may prove to be very useful in wastewater reuse studies, as Clostridium perfringens may be found to have survival characteristics similar to those of viruses or even helminth eggs.

Pathogens

The following pathogenic parameters can only be considered if suitable laboratory facilities and suitably trained staff are available:

- Salmonella spp: Several species of Salmonellae may be present in raw sewage from an urban community in a tropical developing country, including S. typhi (causative agent for typhoid) and many others. It is estimated that a count of 7000 Salmonellae/litre is typical in a tropical urban sewage with similar numbers of Shigellae, and perhaps 1000 Vibrio cholera/litre. Both Shigella spp and V. cholera are more rapidly killed in the environment, so if removal of Salmonellae can be achieved, then the majority of other bacterial pathogens will also have been removed.

- Enteroviruses: May give rise to severe diseases, such as Poliomyelitis and Meningitis, or to a range of minor illnesses such as respiratory infections. Although there is no strong epidemiological evidence for the spread of these diseases via sewage irrigation systems, there is some risk and it is desirable to know to what extent viruses are removed by existing and new treatment processes, especially under tropical conditions. Virus counts can only be undertaken in a dedicated laboratory, as the cell culture techniques required are very susceptible to bacterial and fungal contamination.

- Rotaviruses: These viruses are known to cause gastro-intestinal problems and, though usually present in lower numbers than enterovirusesin sewage, they are known to be more persistent, so it is necessary to establish their survival characteristics relative to enteroviruses and relative to the indicator organisms in wastewaters. It has been claimed that the removal of viruses in wastewater treatment occurs in parallel with the removal of suspended solids, as most virus particles are solids-associated. Hence, the measurement of suspended solids in treated effluents should be carried out as a matter of routine.

- Intestinal Nematodes: It is known that nematode infections, in particular from the round-worm Ascaris lumbricoides, can be spread by effluent reuse practices. The eggs of A. lumbricoides are fairly large (45-70 m m x 35-50 m m) and several techniques for enumeration of nematodes have been developed.

Parameters of Agricultural Significance

The quality of irrigation water is of particular importance in arid zones where extremes of temperature and low relative humidity result in high rates of evaporation, with consequent deposition of salt which tends to accumulate in the soil profile. The physical and mechanical properties of the soil, such as dispersion of particles, stability of aggregates, soil structure and permeability, are very sensitive to the type of exchangeable ions present in irrigation water. Thus, when effluent use is being planned, several factors related to soil properties must be taken into consideration.

Another aspect of agricultural concern is the effect of dissolved solids (TDS) in the irrigation water on the growth of plants. Dissolved salts increase the osmotic potential of soil water and an increase in osmotic pressure of the soil solution increases the amount of energy which plants must expend to take up water from the soil. As a result, respiration is increased and the growth and yield of most plants decline progressively as osmotic pressure increases. Although most plants respond to salinity as a function of the total osmotic potential of soil water, some plants are susceptible to specific ion toxicity.

Many of the ions which are harmless or even beneficial at relatively low concentrations may become toxic to plants at high concentration, either through direct interference with metabolic processes or through indirect effects on other nutrients, which might be rendered inaccessible. Morishita (1985) has reported that irrigation with nitrogen-enriched polluted water can supply a considerable excess of nutrient nitrogen to growing rice plants and can result in a significant yield loss of rice through lodging, failure to ripen and increased susceptibility to pests and diseases as a result of over-luxuriant growth. He further reported that non-polluted soil, having around 0.4 and 0.5 ppm cadmium, may produce about 0.08 ppm Cd in brown rice, while only a little increase up to 0.82, 1.25 or 2.1 ppm of soil Cd has the potential to produce heavily polluted brown rice with 1.0 ppm Cd.

Important agricultural water quality parameters include a number of specific properties of water that are relevant in relation to the yield and quality crops, maintenance of soil productivity and protection of the environment. These parameters mainly consist of certain physical and chemical characteristics of the water. Table below presents a list of some of the important physical and chemical characteristics that are used in the evaluation of agricultural water quality. The primary wastewater quality parameters of importance from an agricultural viewpoint are:

Table: Parameters Used in the Evaluation of Agricultural Water Quality.

Parameters		Symbol	Unit
Physical			
Total dissolved solids		TDS	mg/l
Electrical conductivity		Ec_w	dS/m[1]
Temperature		T	°C
Colour/Turbidity			NTU/JTU[2]
Hardness			mg equiv. $CaCO_3$/l
Sediments			g/l
Chemical			
Acidity/Basicity		pH	
Type and concentration of anions and cations:			
	Calcium	Ca^{++}	me/l[3]
	Magnesium	Mg^{++}	me/l
	Sodium	Na^+	me/l
	Carbonate	CO_3^{--}	me/l
	Bicarbonate	HCO_3^-	me/l
	Chloride	Cl^-	me/l
	Sulphate	SO_4^{--}	me/l
Sodium adsorption ratio		SAR	
Boron		B	mg/l[4]
Trace metals			mg/l
Heavy metals			mg/l
Nitrate-Nitrogen		NO_3-N	mg/l
Phosphate Phosphorus		PO_4-P	mg/l
Potassium		K	mg/l

Total Salt Concentration

Total salt concentration (for all practical purposes, the total dissolved solids) is one of the most important agricultural water quality parameters. This is because the salinity of the soil water is related to, and often determined by, the salinity of the irrigation water. Accordingly, plant growth, crop yield and quality of produce are affected by the total dissolved salts in the irrigation water. Equally, the rate of accumulation of salts in the soil, or soil salinization, is also directly affected by the salinity of the irrigation water. Total salt concentration is expressed in milligrams per litre (mg/l) or parts per million (ppm).

Electrical Conductivity

Electrical conductivity is widely used to indicate the total ionized constituents of water. It is directly related to the sum of the cations (or anions), as determined chemically and is closely correlated, in general, with the total salt concentration. Electrical conductivity is a rapid and reasonably precise determination and values are always expressed at a standard temperature of 25°C to enable comparison of readings taken under varying climatic conditions. It should be noted that the electrical conductivity of solutions increases approximately 2 percent per °C increase in temperature. In this publication, the symbol EC_w, is used to represent the electrical conductivity of irrigation water and the symbol EC_e is used to designate the electrical conductivity of the soil saturation extract. The unit of electrical conductivity is deciSiemen per metre (dS/m).

Sodium Adsorption Ratio

Sodium is an unique cation because of its effect on soil. When present in the soil in exchangeable form, it causes adverse physico-chemical changes in the soil, particularly to soil structure. It has the ability to disperse soil, when present above a certain threshold value, relative to the concentration of total dissolved salts. Dispersion of soils results in reduced infiltration rates of water and air into the soil. When dried, dispersed soil forms crusts which are hard to till and interfere with germination and seedling emergence. Irrigation water could be a source of excess sodium in the soil solution and hence it should be evaluated for this hazard.

The most reliable index of the sodium hazard of irrigation water is the sodium adsorption ration, SAR. The sodium adsorption ratio is defined by the formula:

$$SAR = \frac{Na}{\sqrt{\frac{Ca + Mg}{2}}}$$

where the ionic concentrations are expressed in me/l.

A nomogram for determining the SAR value of irrigation water is presented in Figure. An exchangeable sodium percentage (ESP) scale is included in the nomogram to estimate the ESP value of the soil that is at equilibrium with the irrigation water. Using the nomogram, it is possible to estimate the ESP value of a soil that is at equilibrium with irrigation water of a known SAR value. Under field conditions, the actual ESP may be slightly higher than the estimated equilibrium value because the total salt concentration of the soil solution is increased by evaporation and plant trans-piration, which results in a higher SAR and a corres-pondingly higher ESP value.

It should also be noted that the SAR from Eq $SAR = \frac{Na}{\sqrt{\frac{Ca + Mg}{2}}}$ does not take into account changes in calcium ion concentration in the soil water due to changes in solubility of calcium resulting from precipitation or dissolution during or following an irrigation. However, the SAR calculated according to Eq $SAR = \frac{Na}{\sqrt{\frac{Ca + Mg}{2}}}$ is considered an acceptable evaluation procedure for most of the irrigation waters encountered in agriculture. If significant precipitation or dissolution of calcium due to the effect of carbon dioxide (CO_2), bicarbonate (HCO_3^-) and total salinity (EC_w) is suspected, an alternative procedure for calculating an Adjusted Sodium Adsorption Ratio, SAR_{adj}, can be used.

Toxic Ions

Irrigation water that contains certain ions at concentrations above threshold values can cause plant toxicity problems. Toxicity normally results in impaired growth, reduced yield, changes in the morphology of the plant and even its death. The degree of damage depends on the crop, its stage of growth, the concentration of the toxic ion, climate and soil conditions.

The most common phytotoxic ions that may be present in municipal sewage and treated effluents in concentrations such as to cause toxicity are: boron (B), chloride (Cl) and sodium (Na). Hence, the concentration of these ions will have to be determined to assess the suitability of waste-water quality for use in agriculture.

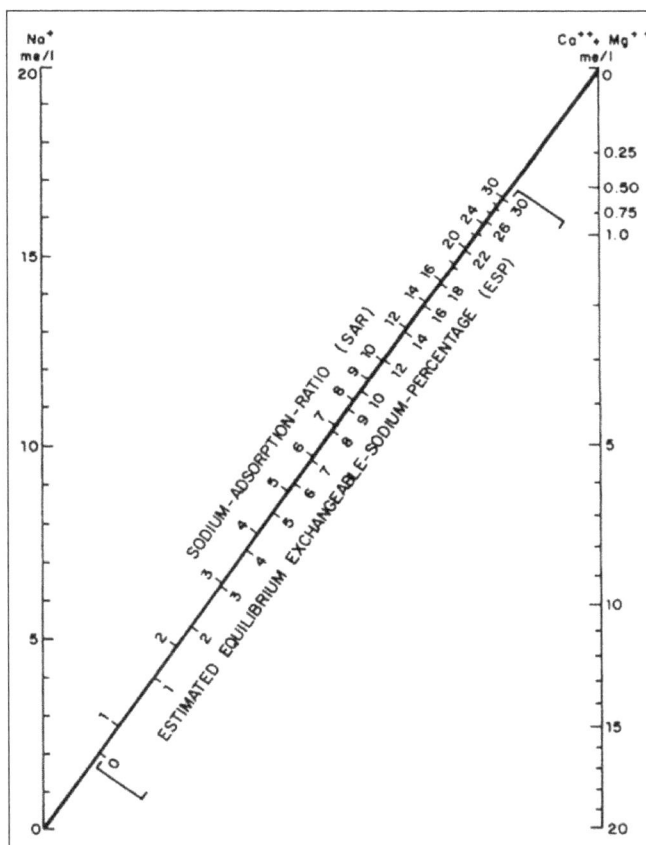

A nomogram for determining sodium adsorption ratio.

Trace Elements and Heavy Metals

A number of elements are normally present in relatively low concentrations, usually less than a few mg/l, in conventional irrigation waters and are called trace elements. They are not normally included in routine analysis of regular irrigation water, but attention should be paid to them when using sewage effluents, particularly if contamination with industrial wastewater discharges is suspected. These include Aluminium (Al), Beryllium (Be), Cobalt (Co), Fluoride (F), Iron (Fe), Lithium (Li), Manganese (Mn), Molybdenum (Mo), Selenium (Se), Tin (Sn), Titanium (Ti), Tungsten (W) and Vanadium (V). Heavy metals are a special group of trace elements which have been shown to create definite health hazards when taken up by plants. Under this group are included, Arsenic (As), Cadmium (Cd), Chromium (Cr), Copper (Cu), Lead (Pb), Mercury (Hg) and Zinc (Zn). These are called heavy metals because in their metallic form, their densities are greaterthan 4g/cc.

pH

pH is an indicator of the acidity or basicity of water but is seldom a problem by itself. The normal pH range for irrigation water is from 6.5 to 8.4; pH values outside this range are a good warning that the water is abnormal in quality. Normally, pH is a routine measurement in irrigation water quality assessment.

Laboratory Wastewater Tests

Industrial, institutional and commercial entities have been required to continually improve the quality of their process wastewater effluent discharges.

At the same time, population and production increases have increased water use, creating a corresponding rise in wastewater quantity. This increased water use and process wastewater generation requires more efficient removal of by-products and pollutants that allows for effluent discharge within established environmental regulatory limits.

The determination of wastewater quality set forth in environmental permits has been established since the 1970s in a series of laboratory tests focused on four major categories:

- Organics: A determination of the concentration of carbon-based (i.e., organic) compounds aimed at establishing the relative "strength" of wastewater (e.g., Biochemical Oxygen Demand (BOD), Chemical Oxygen Demand (COD), Total Organic Carbon (TOC), and Oil and Grease (O&G)).

- Solids: A measurement of the concentration of particulate solids that can dissolve or suspend in wastewater (e.g., Total Solids (TS), Total Suspended Solids (TSS), Total Dissolved Solids (TDS), Total Volatile Solids (TVS), and Total Fixed Solids (TFS)).

- Nutrients: A measurement of the concentration of targeted nutrients (e.g., nitrogen and phosphorus) that can contribute to the acceleration of eutrophication (i.e., the natural aging of water bodies),

- Physical Properties and Other Impact Parameters: Analytical tests designed to measure a varied group of constituents directly impact wastewater treatability (e.g., temperature, color, pH, turbidity, odor).

Although wastewater analytical tests are often separated into categories, it is important to understand that these tests are not independent of each other (Figure). In other words, a contaminant identified by one test in one category can also be identified in another test in a separate category. For example, the organics in a wastewater sample represented by BOD will also be represented in the spectrum of solids, either as suspended (TSS) or dissolved (TDS) particulates. For most people a complete understanding of the standard methods required to accurately complete critical wastewater analytical tests is not necessary. However, a fundamental understanding of the theory behind and working knowledge of the basic procedures used for each test, and the answers to commonly asked questions about each test can be a valuable tool for anyone involved in generating, monitoring, treating or discharging process wastewater.

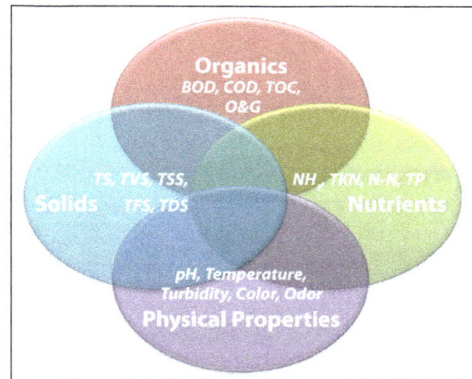

Organics (BOD, COD, TOC, O and G)

Analytical tests aimed at establishing the concentration (typically in mg/L or ppm) of organic (i.e., carbon-containing) matter have traditionally been used to determine the relative "strength" of a wastewater sample. Today there are four common laboratory tests used to determine the gross amount of organic matter (i.e., concentrations > 1.0 mg/L) in wastewater:

- BOD (Biochemical Oxygen Demand).

- COD (Chemical Oxygen Demand).

- TOC (Total Organic Carbon).

- O&G (Oil and Grease).

Wastewater generated by commercial, industrial and institutional facilities is typically referred to as "high-strength" compared to typical household wastewater. Table below shows the typical concentrations (mg/L) of organics found in untreated domestic wastewater. This table can be used to understand how non-sanitary process wastewater compares to typical domestic wastewater.

Table: Typical Concentrations of Organics in Untreated Domestic Wastewater.

Constituents	Unit	Typical Concentration		
		Low	Medium	High
BOD (biochemical oxygen demand)	mg/L	110	190	350
COD (chemical oxygen demand)	mg/L	250	430	800
TOC (total organic carbon)	mg/L	80	140	260
O&G (oil and grease)	mg/L	50	90	100

Wastewater Analytics Acronyms

Organics:

- BOD Biochemical Oxygen Demand,

- COD Chemical Oxygen Demand,

- TOC Total Organic Carbon,

- O&G Oil and Grease.

Solids:

- TS Total Solids,

- TSS Total Suspended Solids,

- TDS Total Dissolved Solids,

- TVS Total Volatile Solids,

- TFS Total Fixed Solids.

Nutrients:

- NH_3 Ammonia,

- TKN Total Kjeldahl Nitrogen,

- N-N Nitrite/Nitrate,

- TP Total Phosphorus.

O&G (Oil and Grease)

- O&G consists of a group of related constituents that are of special concern in wastewater treatment due to their unique physical properties and highly concentrated energy content.

- The term O&G (oil and grease) has become the popular term replacing the term FOG (fat, oil and grease), although both terms refer to the same wastewater constituents.

- O&G constituents in wastewater can come from plants and animals (e.g, lard, butter, vegetable oils and fats) as well as petroleum sources (e.g., kerosene, lubricating oils).

- O&G are generally hydrophobic (i.e., "water-hating") and thus have low solubility in wastewater, resulting in relatively low biodegradability by microorganisms.

- O&G becomes more soluble (i.e., more easily dissolved) in wastewater at high temperatures and will form emulsions (i.e., oil-water mixtures) that will often separate back out of wastewater as temperatures become cooler; thus, O&G are notorious for causing sewer collection system problems (e.g., blockages, pump failures).

- Since O&G adheres to plastic, only glass sample collection containers can be used to collect O&G samples.

O and G Test Procedures

- A clean flask is dried, cooled and weighed.

- A 1L wastewater sample is acidified (typically using hydrochloric or sulfuric acid) to a pH = 2.

- The acidified wastewater sample is then transferred to a 2L separatory funnel.

- 30 mL of the extraction chemical (e.g., n-Hexane) are then added to the funnel and the funnel is shaken vigorously for two minutes.

- The wastewater/extraction chemical layers are allowed to separate in the funnel (the lighter water layer will be on the top and heavier extraction chemical layer will be on the bottom). The bottom layer of extraction chemical is drained into the flask prepared in Step 1.

- Steps 4/5 are repeated twice more to extract O&G.

- The contents of the flask (i.e., the extraction chemical containing O&G) are then heated so that the extraction chemical is distilled into another container.

- The flask (containing the extracted O&G) is reweighed. The original weight of the flask is subtracted and the total O&G weight in mg is calculated. The results provide the O&G concentration in mg/L.

Biochemical Oxygen Demand

BOD is employed to determine the aerobic destructibility of organic substances. As long ago as 1870, Frankland carried out the first BOD measurements, which were very similar to the dilution method in use today.

BOD$_n$ (Biochemical Oxygen Demand after n days) is defined in detail in German Standard DIN 38 409-H51[1] and is associated with certain experimental conditions. It represents the quantity of oxygen which is consumed in the course of aerobic processes of decomposition of organic materials, caused by microorganisms. The BOD therefore provides information on the biologically-convertible proportion of the organic content of a sample of water. This leads to the consideration of these materials in terms of their susceptibility to oxidation by the use of oxygen. BOD is stated in mg/l of oxygen and is usually measured within a period of 5 days (BOD$_5$).

Principle

The self-cleansing ability of water is based on the activity of microorganisms[2] which are present in practically every area of life as a mixed population[3]. They feed on salts and organic compounds such as sugar, cellulose and convertible synthetic substances, which they consume in the presence of oxygen (O^2) - that is, biochemical oxidation occurs and the products are partially or completely broken down.

The expression "total decomposition of organic materials" (C_{org}.) is taken to mean their oxidation to carbon dioxide (CO_2) and inorganic salts (mineralisation), as covered by expression:

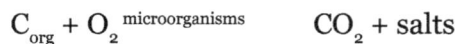

$$C_{org} + O_2 \xrightarrow{\text{microorganisms}} CO_2 + \text{salts}$$

The conditions for determining BOD are so defined that the quantity alone of the organic materials contained in a sample limits the growth of microorganisms - i.e., the greater the number of organic nutrients present, the greater the material-exchange activity of the bacteria and, therefore, the resultant BOD.

The measured consumption of O_2 is the result of microbic conversion. In extremely anaerobic water sources, oxidation processes can also be detected in inorganic materials (e.g., Fe^{2+} → Fe^{3+}).

BOD$_5$ and BOD Graph

Determining BOD values after five days (BOD$_5$) has been adopted as a compromise between a short test-period and the detection of a practically complete biological breakdown of organic materials. With domestic effluents, at 20 °C a complete degradation (= 100 % BOD) is achieved only after 20 days (BOD$_{20}$); however, after only 5 days, 70 % of the biologically convertible substances are broken down.

Between the BOD$_5$ and BOD$_{20}$ values, the following pattern applies: for a given interval of time, the same proportion of residual BOD by volume is broken down. At 20°C, the daily degradation capacity is 20,6 % of the relevant residual BOD (Habeck-Tropfke, 1992) which means that, in ideal conditions and with a total BOD (BOD$_{20}$) of 100 mg/l, after one day there would be a BOD value of 20,6 mg/L. This kinetic represents a first-order reaction. From this relationship it is possible to use the BOD$_5$ reading to estimate values from BOD$_1$ to BOD$_{20}$ for domestic effluents. Further, it is possible to establish a relationship with BOD values taken at other temperatures. The advantages of taking measurements after five days are, therefore, on the one hand, the short amount of time required for the analysis and, on the other, the fact that extrapolation can be carried out.

However, in some countries a testing period of 7 days is performed. A so-called BOD-Graph can be drawn up, in which BOD values are shown in graphical form over a given period of time. A slight deviation from the kinetic described above occurs, caused by fluctuations in the rate of breakdown.

In their natural state, the microorganisms first consume the most easily disintegrated material completely, before they attack the next C-source. The transfer to the subsequent substratum is linked to an adaptation-phase and this causes a (generally short-term) reversal of material conversion activity (diauxia). Further considerations are based on the possibility that, under given standard conditions, adaptations4 of the inoculated microorganisms, caused by their environment, are cancelled out within the 5-day test period, so that comparable BOD values can nevertheless be obtained.

Interpretation

On the basis of the BOD_n value, assertions may be made both with regard to the characteristics of a water-source and the biological activity of the incubated microflora. For example, where a heavy load is placed on a water source, the water may become anaerobic as a result of a lack of oxygen when effluent with a high oxygen demand (consuming a great deal of oxygen) is introduced, and may "tip over". In another case, the biological capacity of a sewage treatment plant can be tested by comparing the BOD value of a known, control solution with the BOD derived from a microbiosphere being present in the treatment plant.

In general, the following assertions may be made:

- A high BOD indicates a high content of easily degradable, organic material in the sample.

- Alow BOD indicates a low volume of organic materials, substances which are difficult to break down or other measuring problems.

- The shape of the BOD graph shows what further information may be gained from the measurements (conformance with the measurement range; problems; pattern of decomposition).

BOD values are generally determined and evaluated in association with other parameters (e.g., COD, DOC, POC, TOC) and this makes them more useful in formulating predictions. For example, if we consider a comparison of the measured BOD value with the COD value:

- A small difference indicates that a large proportion of the organic materials can easily be degraded.

- A large difference indicates either that the organic loading cannot be easily broken down, or that a problem is present.

BOD detects only the destructible proportion of organic substances and as a general principle is therefore lower than the COD value, which also includes inorganic materials and those materials which cannot be biologically oxidised.

Problems

Biodegradation of materials present in water occurs in two phases. The breakdown of carbons begins practically immediately. In the second phase, nitrification occurs, which also involves the consumption of oxygen. There are two groups of nitrifying bacteria, which catalyse the synthesis in a close relationship. As shown in expression the first group oxidises ammonium (NH_4^+) to nitrite (NO_2), which represents the substratum for the second group, which creates nitrate (NO_3):

$$NH_4^+ \qquad\qquad NO_2 - \qquad\qquad NO_3 -$$

This conversion requires 4.57 mg/L of oxygen per mg of NH_4^+ and has a significant effect on the BOD, which is intended only to detect the consumption of oxygen for carbon oxidation (CBOD). Generally speaking, nitrification sets in after 10 days but may occur within 5 days in samples which are not heavily loaded5. To repress this undesirable occurrence, nitrification inhibitors, such as NAllylthiourea (ATH) or 2-chloro-6-(trichloro methyl) pyridine (CTMP) can be added. If it is desired to analyse oxygen consumption of the nitrification in itself (N-BOD) a comparison can be made of samples both with and without nitrification inhibitors.

In contrast to this, inhibitors or toxic substances can reduce biological activity in water, or even kill off the organisms completely. Even so, it may be of interest to determine the BOD of a sample of this kind. This can be achieved with a dilute solution, in which substances such as metabolic poisons are below the concentration at which they would have an effect on aerobic decomposition.

Procedures

There are two standard procedures, which are equivalent to each other under given conditions: the first is the dilution method to DIN 38409-H51 and gives the so-called dilution BOD_n over a period of n days. This is in most cases an officially-approved procedure. It is cumbersome and includes an optimization for test conditions required for the decomposition process at 20 °C. The oxygen content is measured directly in the water itself.

The more simple procedure is to measure oxygen depletion $Z_{S(n)}$ during a period of n days (as DIN 38409-H52), and this method is used for internal monitoring purposes. The respirometric method (manometer) is based on the fact that the oxygen converted to carbon dioxide is removed from the gaseous spectrum via the sample by means of a CO_2 adsorbent, usually potassium hydroxide (KOH). The resultant pressure drop can be measured in the closed system and is proportional to the volume of the consumed oxygen. The so called oxygen depletion value ($Z_{S(n)}$) can be interpreted as a BOD_n value where:

- Only the concentration of organic substances limits oxygen consumption in the samples.

- The investigation takes place at 20 ± 1 °C.

In the past, pressure measurements were usually carried out with a mercury manometer. However, this can be done more simply and reliably, using an electronic BOD-OxiDirect from Lovibond.

Following measurement, the results are stored and can be called up to the display at any time in mg/L BOD. In particular, this eliminates the use of mercury, which can be dangerous to health.

Further Applications for the BOD Method

BOD is a criterion for aerobic, biological decomposition. A special application is found in water analysis as the BOD_n value. Variations of the method can lead to other possible applications, such as the following:

- Checks on the aerobic decomposition of environmental chemicals (e.g., biodegradability to OECD 301) within 28 days.

- Determining the respiration of soils, sludge, sediment, garbage and liquids.

- Toxicity tests in soils, sludge, sediment, garbage and liquids.

- Bio-activity determination in various environ mental compartments.

- Performance checks in a sewage treatment plant.

- Determining the respiration rates of living beings (e.g., R/Q-value (respiration quotient), investigations of stress).

- Oxygen consumption of cell cultures under the influence of various tests in the medicinal & pharmaceutical industry.

Typical Values

Examples of typical BOD_5 values for natural and anthropogenous water sources shown in Table 1 are given for purposes of general guidance:

Table: Typical BOD5 Values.

Water	BOD_5 [mg/I]	Effluent/ Outfall	BOD_5 [mg/I]
GKI. IV	>15	Blood*	160.000 - 210.000
GKI. IV/III	20-10	Liquid manure	7.000-18.000
GKI. III	13-7	Whey*	45
GKI. III/II	10-5	Mosel River (D)	1,0 - 5,1
GKI. II	6-2	Nahe River (D)	<1,0 -5,1
GKI. II/I	2-1	Rhein River (D)	<1,0 - 1,9
GKI. I	≤1	SP-inlet	300-350
Drinking water	<1	S-outflow	<25

Chemical Oxygen Demand

Chemical oxygen demand is the amount of oxygen needed to oxidize the organic matter present in water. Chemical oxygen demand testing is used to determine the amount of oxidation that will occur and the amount of organic matter in a water sample. Chemical oxygen demand testing is also used to determine the amount of inorganic chemicals in a sample.

Chemical oxygen demand testing is typically performed using a strong oxidizing chemical. Organic matter is oxidized into carbon dioxide and water in an acidic condition. The quantity of organic matter or the demand of oxygen is calculated by determining how much oxidizing chemical was consumed during the test.

Chemical oxygen demand tests are typically performed on wastewater. The pollution level is calculated by measuring the amount of organic matter in the water. Water with too much organic material can have a negative effect on the environment in which the wastewater is discharged.

Chemical oxygen demand is similar to biochemical oxygen demand in that they are both used to calculate the oxygen demand of a water sample. The difference between the two is that chemical oxygen demand measures everything that can be oxidized, whereas biochemical oxygen demand only measures the oxygen demanded by organisms.

Total Organic Carbon

Organic carbon compounds vary greatly. In fact, one of the first lessons in most introductory Organic Chemistry courses explains that the number of possible carbon compounds is virtually infinite due to carbon's ability to form long, chain-like molecules. While chromatographic methods like gas chromatography (GC) or high-performance liquid chromatography (HPLC) are able to make quantitative determinations for specific compounds, the user must first know which specific compounds to look for.

Total organic carbon (TOC) is a non-specific test, which means TOC will not determine which particular compounds are present (most samples are complex mixtures which contain thousands of different organic carbon compounds). Instead, TOC will inform the user of the sum of all organic carbon within those compounds.

The reasons for measuring TOC vary across industries, but generally fall into two categories: process control, or regulatory compulsion. Some of the most common TOC measurement applications include:

- Municipal Drinking Water: Organic carbon reacts with disinfection chemicals such as chlorine and forms disinfection byproducts (DBP), which may be carcinogenic. Reducing organic carbon prior to disinfection can significantly decrease harmful DBP exposure for the public.

- Municipal Waste Water: Monitoring organic carbon of influent facilitates process controls for maximizing plant efficiency, while monitoring effluent is often a requirement for discharging into surface waters.

- Industrial Waste Water: Industries which discharge liquid waste into a surface water body are required to monitor TOC.

- Power Plants: Limiting potential sources of corrosive compounds can prevent costly damage to expensive equipment.

- Pharmaceutical Manufacturers: Water is the most commonly used ingredient used to produce drugs. Regulations limit the concentration of organic carbon to prevent harmful bacteria from growing.

- Electronics Manufacturers: Ultra-pure water is used in the manufacture of microprocessors and computer chips. As processors and circuits become smaller and smaller the water must be kept incredibly clean to prevent microscopic damage to these miniature circuits.

TOC Detection Methods

Several methods exist for measuring TOC, however each method has two common objectives: 1) Oxidize organic carbon to carbon dioxide, and 2) measure the carbon dioxide generated.

Common oxidation methods include chemical agents (such as persulfate), combustion (usually aided by a catalyst), exposure to ionizing radiation (such as ultra violet light), exposure to heat, or some combination of these methods.

There are fewer options for detecting carbon dioxide, two common methods are conductivity and non-dispersive infra-red (NDIR). Conductivity based detection methods work by sensing an increase in ion concentration which is attributed to the increased presence of bicarbonate and carbonate ions created from the oxidation of organic compounds. Non-dispersive Infra-red detectors measure carbon dioxide by determining the amount of infra-red light absorbed across a known distance.

Preventing Damage to the Instrument

Two common pitfalls that can damage TOC measurement instruments or produce erroneous measurement results include sample overload (running a sample which far exceeds the maximum analyte specification), and carryover (contamination from a previous sample).

Overload conditions are common when running unknown samples. Depending on the measurement technology used this condition can cause costly damage to an instrument. For example, a combustion-type instrument which uses platinum catalysts can very easily ruin the catalysts and require expensive replacements. Membrane-based TOC measurement instruments can also coat the surface of the membrane with organic carbon compounds from a high-concentration unknown sample. Such an event will leave the instrument out of operation while awaiting service.

Carryover results from residual sample left from a previous measurement. It is most often observed when multiple replicates of a sample are measured, and a high concentration sample is followed by a low concentration sample. The following equation calculates carryover as a percentage of the difference between the two sample concentrations:

Methods of Calculating TOC

Inorganic carbon is bound only to oxygen, as in carbon dioxide, bicarbonate, or carbonate (for example: limestone is calcium carbonate which is a form of inorganic carbon). Organic carbon can be bound to a variety of other elements such as hydrogen, nitrogen, or other carbon atoms.

Other forms of carbon include purgeable versus non-purgeable carbon. Volatile organic compounds have a low boiling point, and can be purged from a solution by bubbling gas through a sample.

The following abbreviations are commonly used to describe various forms of carbon when measuring TOC:

- TC: Total Carbon,
- TOC: Total Organic Carbon,

- TIC: Total Inorganic Carbon,

- POC: Purgeable Organic Carbon (also called VOC or Volatile Organic Carbon),

- NPOC: Non-purgeable Organic Carbon,

- Calculating TOC can be done by subtracting the TIC from the TC. This method is described by the equation,

- TC − TIC = TOC.

This method works well when there is a large difference between TC and TIC; however when TIC values are high the difference method can produce very erratic results because the margin of error for both the TC measurement and the TIC measurement must be added together.

In many TOC measurement applications it is reasonable to assume that the contribution of POC to the overall TOC value is negligible, and therefore the following approximation is used.

NPOC ≈ TOC

This approximation is good for drinking water, where the largest contribution of organic carbon comes from humic acids which are non-volatile, high molecular weight compounds. Ultra-pure applications such as pharmaceutical, power, and semiconductor manufacturing also should expect to have negligible concentrations of POC present in the sample.

The NPOC methods usually employ NDIR measurement technology, which generates a signal that is recorded over time. When the signal is graphed two peaks are prominent. The first peak results from inorganic carbon (dissolved CO_2 already present in the sample). The second peak results from the organic carbon which is oxidized to CO_2. The graph below illustrates an NPOC measurement cycle.

Adsorbable Organic Halides

Adsorbable organic halides (AOX) is an organic sum parameter comprising such organics that contain chlorine, bromine or iodine (not fluorine) atoms and are adsorbable to activated carbon. For AOX determination a particular volume of the wastewater sample is agitated sufficiently long with powdered activated carbon. Subsequently the activated carbon is separated by filtration using a membrane filter which retains the activated carbon (adsorption can also be executed in small activated carbon columns which are treated - after adsorption has been completed - in the same way as the loaded activated carbon removed by filtration). Then the membrane filter is incinerated together with the activated carbon in a stream of pure oxygen at temperatures around 900°C. The halogen atoms originally bound in organics adsorbed to the activated carbon form HCl, HBr, or HI, resp., which are contained in the exhaust gas of the incineration furnace and can be absorbed e.g. in acetic acid. Microcoulometric titration, an electrochemical quantification method, analyses chloride, bromide, or iodide, resp., of these acids. Bromide and iodide are calculated as chloride equivalents (one mol bromide or iodide is looked at as one mol chloride and is calculated as chloride mass), and the final chloride mass determined is related to the volume of the wastewater sample which had been subdued to activated carbon adsorption. The result is mg AOX (chloride)/l wastewater.

In the AOX analysis procedure, artefacts can easily be produced: First, also inorganic chloride adsorbs to a certain amount to activated carbon. This adsorbed inorganic chloride will also be detected e.g. by microcoulometric analysis of the incineration off-gass and may result in the so-called "chloride error". Secondly, in wastewaters with high TOC mainly represented by non-halogenated organic compounds a competition of halogenated and non-halogenated organic compounds for adsorption sites on the activated carbon occurs leading to a very low extent of halogenated organic molecules being adsorbed. This can be prevented by dilution of the wastewater sample. However, by dilution also the AOX is diluted which is disadvantageous if the AOX content of the sample is decreased to be below the detection limit of the method. AOX analyses must be performed in laboratory rooms where no halogenated organic solvents are used at all, because these volatiles would also adsorb on the activated carbon during the AOX procedure. In recent years, AOX analyses had been performed in a laboratory where a thermostatized chamber was located. When there was a leakage in the cooling system of the chamber, some fluorochlorohydrocarbons were volatilized in the laboratory leading to severe analytical errors in AOX determinations.

Other parts of organics contained in wastewaters (usually comprised in TOC or COD) are the organic sum parameters hydrocarbons, phenols, anionic surfactants, neutral surfactants, cationic surfactants etc. Methods for analyzing these organic sum parameters are also given in the "Standard Methods".

Chapter 3

Wastewater Treatment

The process of removing contaminants from wastewater and converting them into reusable water is defined as wastewater treatment. Some of the wastewater treatment systems are grit removal, wastewater chlorination, activated sludge process, etc. The topics elaborated in this chapter will help in gaining a better perspective about these different systems of wastewater treatment.

Wastewater treatment is the removal of impurities from wastewater, or sewage, before they reach aquifers or natural bodies of water such as rivers, lakes, estuaries, and oceans. Since pure water is not found in nature (i.e., outside chemical laboratories), any distinction between clean water and polluted water depends on the type and concentration of impurities found in the water as well as on its intended use. In broad terms, water is said to be polluted when it contains enough impurities to make it unfit for a particular use, such as drinking, swimming, or fishing. Although water quality is affected by natural conditions, the word pollution usually implies human activity as the source of contamination. Water pollution, therefore, is caused primarily by the drainage of contaminated wastewater into surface water or groundwater, and wastewater treatment is a major element of water pollution control.

Sources of Water Pollution

Water pollutants may originate from point sources or from dispersed sources. A point-source pollutant is one that reaches water from a single pipeline or channel, such as a sewage discharge or outfall pipe. Dispersed sources are broad, unconfined areas from which pollutants enter a body of water. Surface runoff from farms, for example, is a dispersed source of pollution, carrying animal wastes, fertilizers, pesticides, and silt into nearby streams. Urban storm water drainage, which may carry sand and other gritty materials, petroleum residues from automobiles, and road deicing chemicals, is also considered a dispersed source because of the many locations at which it enters local streams or lakes. Point-source pollutants are easier to control than dispersed-source pollutants, since they flow to a single location where treatment processes can remove them from the water. Such control is not usually possible over pollutants from dispersed sources, which cause a large part of the overall water pollution problem. Dispersed-source water pollution is best reduced by enforcing proper land-use plans and development standards.

General types of water pollutants include pathogenic organisms, oxygen-demanding wastes, plant nutrients, synthetic organic chemicals, inorganic chemicals, microplastics, sediments, radioactive substances, oil, and heat. Sewage is the primary source of the first three types. Farms and industrial facilities are also sources of some of them. Sediment from eroded topsoil is considered a pollutant because it can damage aquatic ecosystems, and heat (particularly from power-plant cooling water) is considered a pollutant because of the adverse effect it has on dissolved oxygen levels and aquatic life in rivers and lakes.

Sewage Characteristics

Types of Sewage

There are three types of wastewater, or sewage: domestic sewage, industrial sewage, and storm sewage. Domestic sewage carries used water from houses and apartments; it is also called sanitary sewage. Industrial sewage is used water from manufacturing or chemical processes. Storm sewage, or storm water, is runoff from precipitation that is collected in a system of pipes or open channels.

Domestic sewage is slightly more than 99.9 percent water by weight. The rest, less than 0.1 percent, contains a wide variety of dissolved and suspended impurities. Although amounting to a very small fraction of the sewage by weight, the nature of these impurities and the large volumes of sewage in which they are carried make disposal of domestic wastewater a significant technical problem. The principal impurities are putrescible organic materials and plant nutrients, but domestic sewage is also very likely to contain disease-causing microbes. Industrial wastewater usually contains specific and readily identifiable chemical compounds, depending on the nature of the industrial process. Storm sewage carries organic materials, suspended and dissolved solids, and other substances picked up as it travels over the ground.

Principal Pollutants

Organic Material

The amount of putrescible organic material in sewage is indicated by the biochemical oxygen demand, or BOD; the more organic material there is in the sewage, the higher the BOD, which is the amount of oxygen required by microorganisms to decompose the organic substances in sewage. It is among the most important parameters for the design and operation of sewage treatment plants. Industrial sewage may have BOD levels many times that of domestic sewage. The BOD of storm sewage is of particular concern when it is mixed with domestic sewage in combined sewerage systems.

Dissolved oxygen is an important water quality factor for lakes and rivers. The higher the concentration of dissolved oxygen, the better the water quality. When sewage enters a lake or stream, decomposition of the organic materials begins. Oxygen is consumed as microorganisms use it in their metabolism. This can quickly deplete the available oxygen in the water. When the dissolved oxygen levels drop too low, trout and other aquatic species soon perish. In fact, if the oxygen level drops to zero, the water will become septic. Decomposition of organic compounds without oxygen causes the undesirable odours usually associated with septic or putrid conditions.

Suspended Solids

Another important characteristic of sewage is suspended solids. The volume of sludge produced in a treatment plant is directly related to the total suspended solids present in the sewage. Industrial and storm sewage may contain higher concentrations of suspended solids than domestic sewage. The extent to which a treatment plant removes suspended solids, as well as BOD, determines the efficiency of the treatment process.

Plant Nutrients

Domestic sewage contains compounds of nitrogen and phosphorus, two elements that are basic nutrients essential for the growth of plants. In lakes, excessive amounts of nitrates and phosphates can cause the rapid growth of algae. Algal blooms, often caused by sewage discharges, accelerate the natural aging of lakes in a process called eutrophication.

Microbes

Domestic sewage contains many millions of microorganisms per gallon. Most are coliform bacteria from the human intestinal tract, and domestic sewage is also likely to carry other microbes. Coliforms are used as indicators of sewage pollution. A high coliform count usually indicates recent sewage pollution.

Sewerage Systems

A sewerage system, or wastewater collection system, is a network of pipes, pumping stations, and appurtenances that convey sewage from its points of origin to a point of treatment and disposal.

Combined Systems

Systems that carry a mixture of both domestic sewage and storm sewage are called combined sewers. Combined sewers typically consist of large-diameter pipes or tunnels, because of the large volumes of storm water that must be carried during wet-weather periods. They are very common in older cities but are no longer designed and built as part of new sewerage facilities. Because wastewater treatment plants cannot handle large volumes of storm water, sewage must bypass the treatment plants during wet weather and be discharged directly into the receiving water. These combined sewer overflows, containing untreated domestic sewage, cause recurring water pollution problems and are very troublesome sources of pollution.

In some large cities the combined sewer overflow problem has been reduced by diverting the first flush of combined sewage into a large basin or underground tunnel. After temporary storage, it can be treated by settling and disinfection before being discharged into a receiving body of water, or it can be treated in a nearby wastewater treatment plant at a rate that will not overload the facility. Another method for controlling combined sewage involves the use of swirl concentrators. These direct sewage through cylindrically shaped devices that create a vortex, or whirlpool, effect. The vortex helps concentrate impurities in a much smaller volume of water for treatment.

Separate Systems

New wastewater collection facilities are designed as separate systems, carrying either domestic sewage or storm sewage but not both. Storm sewers usually carry surface runoff to a point of disposal in a stream or river. Small detention basins may be built as part of the system, storing storm water temporarily and reducing the magnitude of the peak flow rate. Sanitary sewers, on the other hand, carry domestic wastewater to a sewage treatment plant. Pretreated industrial wastewater may be allowed into municipal sanitary sewerage systems, but storm water is excluded.

Storm sewers are usually built with sections of reinforced concrete pipe. Corrugated metal pipes may be used in some cases. Storm water inlets or catch basins are located at suitable intervals in a street right-of-way or in easements across private property. The pipelines are usually located to allow downhill gravity flow to a nearby stream or to a detention basin. Storm water pumping stations are avoided, if possible, because of the very large pump capacities that would be needed to handle the intermittent flows.

A sanitary sewerage system includes laterals, submains, and interceptors. Except for individual house connections, laterals are the smallest sewers in the network. They usually are not less than 200 mm (8 inches) in diameter and carry sewage by gravity into larger submains, or collector sewers. The collector sewers tie in to a main interceptor, or trunk line, which carries the sewage to a treatment plant. Interceptors are usually built with precast sections of reinforced concrete pipe, up to 5 metres (15 feet) in diameter. Other materials used for sanitary sewers include vitrified clay, asbestos cement, plastic, steel, or ductile iron. The use of plastic for laterals is increasing because of its lightness and ease of installation. Iron and steel pipes are used for force mains or in pumping stations. Force mains are pipelines that carry sewage under pressure when it must be pumped.

Alternative Systems

Sometimes the cost of conventional gravity sewers can be prohibitively high because of low population densities or site conditions such as a high water table or bedrock. Three alternative wastewater collection systems that may be used under these circumstances include small-diameter gravity sewers, pressure sewers, and vacuum sewers.

In small-diameter gravity systems, septic tanks are first used to remove settleable and floating solids from the wastewater from each house before it flows into a network of collector mains (typically 100 mm, or 4 inches, in diameter); these systems are most suitable for small rural communities. Because they do not carry grease, grit and sewage solids, the pipes can be of smaller diameter and placed at reduced slopes or gradients to minimize trench excavation costs. Pressure sewers are best used in flat areas or where expensive rock excavation would be required. Grinder pumps discharge wastewater from each home into the main pressure sewer, which can follow the slope of the ground. In a vacuum sewerage system, sewage from one or more buildings flows by gravity into a sump or tank from which it is pulled out by vacuum pumps located at a central vacuum station and then flows into a collection tank. From the vacuum collection tank the sewage is pumped to a treatment plant.

Pumps

Pumping stations are built when sewage must be raised from a low point to a point of higher elevation or where the topography prevents downhill gravity flow. Special nonclogging pumps are available to handle raw sewage. They are installed in structures called lift stations. There are two basic types of lift stations: dry well and wet well. A wet-well installation has only one chamber or tank to receive and hold the sewage until it is pumped out. Specially designed submersible pumps and motors can be located at the bottom of the chamber, completely below the water level. Dry-well installations have two separate chambers, one to receive the wastewater and one to enclose and protect the pumps and controls. The protective dry chamber allows easy access for inspection and maintenance. All sewage lift stations, whether of the wet-well or dry-well type, should include at least two pumps. One pump can operate while the other is removed for repair.

Flow Rates

There is a wide variation in sewage flow rates over the course of a day. A sewerage system must accommodate this variation. In most cities domestic sewage flow rates are highest in the morning and evening hours. They are lowest during the middle of the night. Flow quantities depend upon population density, water consumption, and the extent of commercial or industrial activity in the community. The average sewage flow rate is usually about the same as the average water use in the community. In a lateral sewer, short-term peak flow rates can be roughly four times the average flow rate. In a trunk sewer, peak flow rates may be two-and-a-half times the average.

Although sewage flows depend upon residential, commercial, and industrial connections, sewage flow rates potentially can become higher as a result of inflows and infiltration (I&I) into the sanitary sewer system. Inflows correspond to storm water entering sewers from inappropriate connections, such as roof drains, storm drains, downspouts and sump pumps. High amounts of rainwater runoff can reach the sewer system during precipitation and stormflow events or during seasonal spring flooding of rivers inundated with melting ice. Infiltration refers to the groundwater entering sewers via defective or broken pipes. In both these cases, downstream utilities and treatment plants may experience flows higher than anticipated and can become hydraulically overloaded. During such overloads, utilities may ask residents connected to the system to refrain from using dishwashers and washing machines and may even limit toilet flushing and the use of showers in an attempt to lessen the strain. Such I&I issues can be especially severe in old and aging water infrastructures.

Wastewater Treatment and Disposal

The size and capacity of wastewater treatment systems are determined by the estimated volume of sewage generated from residences, businesses, and industries connected to sewer systems as well as the anticipated inflows and infiltration (I&I). The selection of specific on-lot, clustered, or centralized treatment plant configurations depends upon factors such as the number of customers being served, the geographical scenario, site constraints, sewer connections, average and peak flows, influent wastewater characteristics, regulatory effluent limits, technological feasibility, energy consumption, and the operations and maintenance costs involved.

The predominant method of wastewater disposal in large cities and towns is discharge into a body of surface water. Suburban and rural areas rely more on subsurface disposal. In either case, wastewater must be purified or treated to some degree in order to protect both public health and water quality. Suspended particulates and biodegradable organics must be removed to varying extents. Pathogenic bacteria must be destroyed. It may also be necessary to remove nitrates and phosphates (plant nutrients) and to neutralize or remove industrial wastes and toxic chemicals.

The degree to which wastewater must be treated varies, depending on local environmental conditions and governmental standards. Two pertinent types of standards are stream standards and effluent standards. Stream standards, designed to prevent the deterioration of existing water quality, set limits on the amounts of specific pollutants allowed in streams, rivers, and lakes. The limits depend on a classification of the "maximum beneficial use" of the water. Water quality parameters that are regulated by stream standards include dissolved oxygen, coliforms, turbidity, acidity, and toxic substances. Effluent standards, on the other hand, pertain directly to the quality of the treated wastewater discharged from a sewage treatment plant. The factors controlled under these standards usually include biochemical oxygen demand (BOD), suspended solids, acidity, and coliforms.

There are three levels of wastewater treatment: primary, secondary, and tertiary (or advanced). Primary treatment removes about 60 percent of total suspended solids and about 35 percent of BOD; dissolved impurities are not removed. It is usually used as a first step before secondary treatment. Secondary treatment removes more than 85 percent of both suspended solids and BOD. A minimum level of secondary treatment is usually required in the United States and other developed countries. When more than 85 percent of total solids and BOD must be removed, or when dissolved nitrate and phosphate levels must be reduced, tertiary treatment methods are used. Tertiary processes can remove more than 99 percent of all the impurities from sewage, producing an effluent of almost drinking-water quality. Tertiary treatment can be very expensive, often doubling the cost of secondary treatment. It is used only under special circumstances.

For all levels of wastewater treatment, the last step prior to discharge of the sewage effluent into a body of surface water is disinfection, which destroys any remaining pathogens in the effluent and protects public health. Disinfection is usually accomplished by mixing the effluent with chlorine gas or with liquid solutions of hypochlorite chemicals in a contact tank for at least 15 minutes. Because chlorine residuals in the effluent may have adverse effects on aquatic life, an additional chemical may be added to dechlorinate the effluent. Ultraviolet radiation, which can disinfect without leaving any residual in the effluent, is becoming more competitive with chlorine as a wastewater disinfectant.

Primary Treatment

Primary treatment removes material that will either float or readily settle out by gravity. It includes the physical processes of screening, comminution, grit removal, and sedimentation. Screens are made of long, closely spaced, narrow metal bars. They block floating debris such as wood, rags, and other bulky objects that could clog pipes or pumps. In modern plants the screens are cleaned mechanically, and the material is promptly disposed of by burial on the plant grounds. A comminutor may be used to grind and shred debris that passes through the screens. The shredded material is removed later by sedimentation or flotation processes.

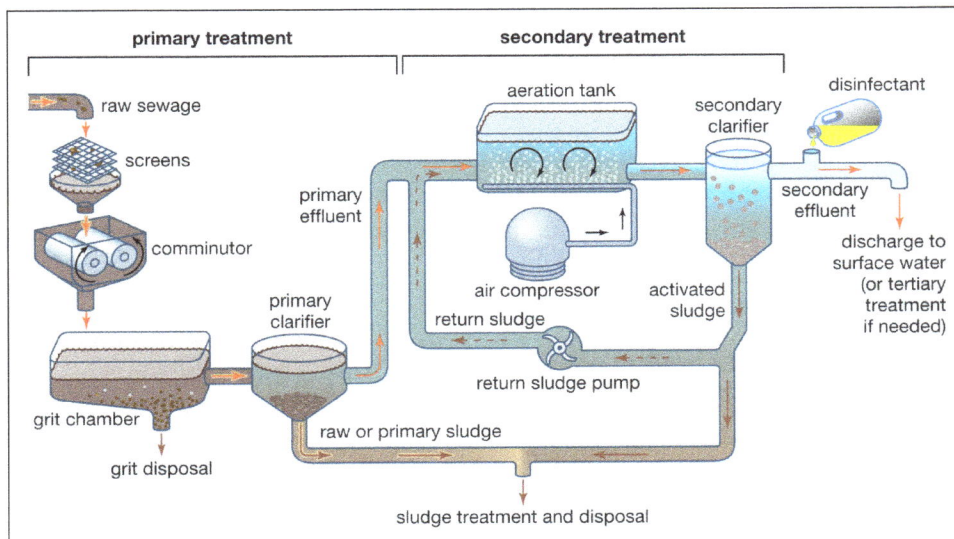

Primary and secondary treatment of sewage, using the activated sludge process.

Grit chambers are long narrow tanks that are designed to slow down the flow so that solids such as sand, coffee grounds, and eggshells will settle out of the water. Grit causes excessive wear and tear on pumps and other plant equipment. Its removal is particularly important in cities with combined sewer systems, which carry a good deal of silt, sand, and gravel that wash off streets or land during a storm.

Suspended solids that pass through screens and grit chambers are removed from the sewage in sedimentation tanks. These tanks, also called primary clarifiers, provide about two hours of detention time for gravity settling to take place. As the sewage flows through them slowly, the solids gradually sink to the bottom. The settled solids—known as raw or primary sludge—are moved along the tank bottom by mechanical scrapers. Sludge is collected in a hopper, where it is pumped out for removal. Mechanical surface-skimming devices remove grease and other floating materials.

Secondary Treatment

Secondary treatment removes the soluble organic matter that escapes primary treatment. It also removes more of the suspended solids. Removal is usually accomplished by biological processes in which microbes consume the organic impurities as food, converting them into carbon dioxide, water, and energy for their own growth and reproduction. The sewage treatment plant provides a suitable environment, albeit of steel and concrete, for this natural biological process. Removal of soluble organic matter at the treatment plant helps to protect the dissolved oxygen balance of a receiving stream, river, or lake.

There are three basic biological treatment methods: the trickling filter, the activated sludge process, and the oxidation pond. A fourth, less common method is the rotating biological contacter.

Trickling Filter

A trickling filter is simply a tank filled with a deep bed of stones. Settled sewage is sprayed continuously over the top of the stones and trickles to the bottom, where it is collected for further treatment. As the wastewater trickles down, bacteria gather and multiply on the stones. The steady flow of sewage over these growths allows the microbes to absorb the dissolved organics, thus lowering the biochemical oxygen demand (BOD) of the sewage. Air circulating upward through the spaces among the stones provides sufficient oxygen for the metabolic processes.

Settling tanks, called secondary clarifiers, follow the trickling filters. These clarifiers remove microbes that are washed off the rocks by the flow of wastewater. Two or more trickling filters may be connected in series, and sewage can be recirculated in order to increase treatment efficiencies.

Activated Sludge

The activated sludge treatment system consists of an aeration tank followed by a secondary clarifier. Settled sewage, mixed with fresh sludge that is recirculated from the secondary clarifier, is introduced into the aeration tank. Compressed air is then injected into the mixture through porous diffusers located at the bottom of the tank. As it bubbles to the surface, the diffused air provides oxygen and a rapid mixing action. Air can also be added by the churning action of mechanical propeller-like mixers located at the tank surface.

Under such oxygenated conditions, microorganisms thrive, forming an active, healthy suspension of biological solids-mostly bacteria-called activated sludge. About six hours of detention is provided in the aeration tank. This gives the microbes enough time to absorb dissolved organics from the sewage, reducing the BOD. The mixture then flows from the aeration tank into the secondary clarifier, where activated sludge settles out by gravity. Clear water is skimmed from the surface of the clarifier, disinfected, and discharged as secondary effluent. The sludge is pumped out from a hopper at the bottom of the tank. About 30 percent of the sludge is recirculated back into the aeration tank, where it is mixed with the primary effluent. This recirculation is a key feature of the activated sludge process. The recycled microbes are well acclimated to the sewage environment and readily metabolize the organic materials in the primary effluent. The remaining 70 percent of the secondary sludge must be treated and disposed of in an acceptable manner.

Variations of the activated sludge process include extended aeration, contact stabilization, and high-purity oxygen aeration. Extended aeration and contact stabilization systems omit the primary settling step. They are efficient for treating small sewage flows from motels, schools, and other relatively isolated wastewater sources. Both of these treatments are usually provided in prefabricated steel tanks called package plants. Oxygen aeration systems mix pure oxygen with activated sludge. A richer concentration of oxygen allows the aeration time to be shortened from six to two hours, reducing the required tank volume.

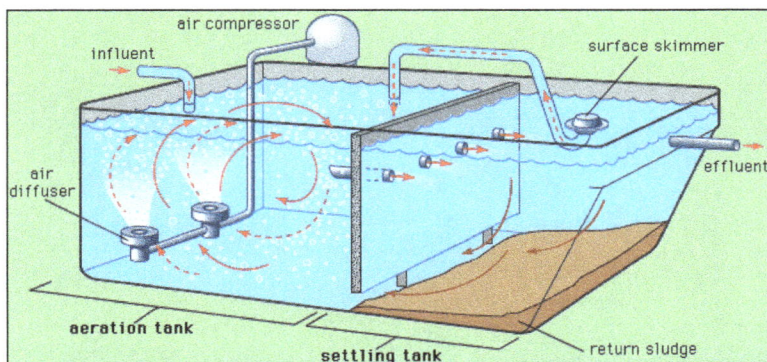

Schematic diagram of a prefabricated package plant for the aeration treatment of small sewage flows.

Oxidation Pond

Oxidation ponds, also called lagoons or stabilization ponds, are large, shallow ponds designed to treat wastewater through the interaction of sunlight, bacteria, and algae. Algae grow using energy from the sun and carbon dioxide and inorganic compounds released by bacteria in water. During the process of photosynthesis, the algae release oxygen needed by aerobic bacteria. Mechanical aerators are sometimes installed to supply yet more oxygen, thereby reducing the required size of the pond. Sludge deposits in the pond must eventually be removed by dredging. Algae remaining in the pond effluent can be removed by filtration or by a combination of chemical treatment and settling.

Rotating Biological Contacter

In this treatment system a series of large plastic disks mounted on a horizontal shaft are partially submerged in primary effluent. As the shaft rotates, the disks are exposed alternately to air and wastewater, allowing a layer of bacteria to grow on the disks and to metabolize the organics in the wastewater.

Tertiary Treatment

When the intended receiving water is very vulnerable to the effects of pollution, secondary effluent may be treated further by several tertiary processes.

Effluent Polishing

For the removal of additional suspended solids and BOD from secondary effluent, effluent polishing is an effective treatment. It is most often accomplished using granular media filters, much like the filters used to purify drinking water. Polishing filters are usually built as prefabricated units, with tanks placed directly above the filters for storing backwash water. Effluent polishing of wastewater may also be achieved using microstrainers of the type used in treating municipal water supplies.

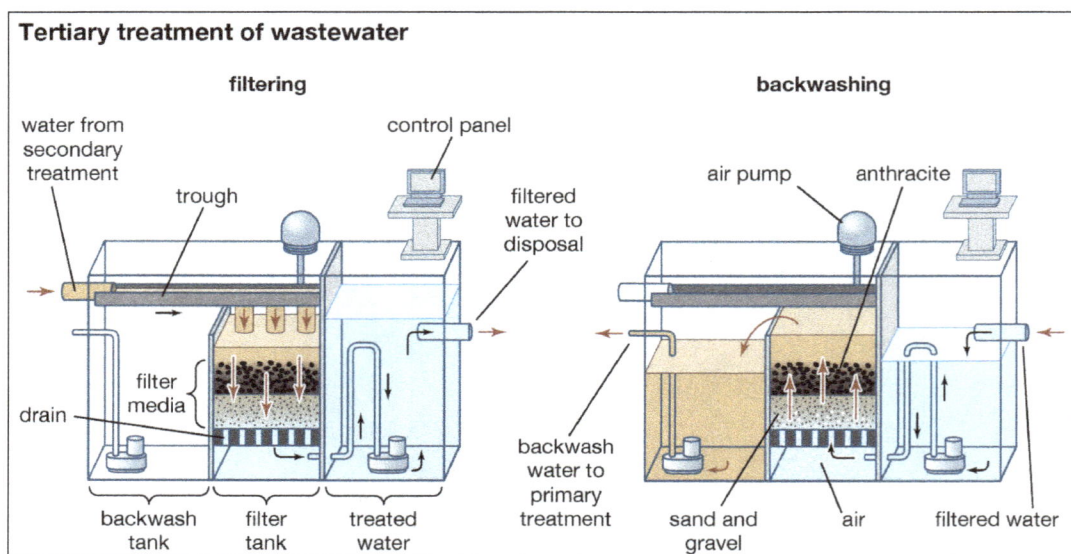

Tertiary treatment of wastewater.

During the filtering step, wastewater from secondary treatment, still containing suspended solids, pours from a trough and percolates through a filter bed made of porous media such as sand, gravel, and anthracite. The filtered water is then piped away for disposal. In the backwashing step, entrained solids are periodically flushed from the filter media by pumping filtered water back through the assembly. The backwash water, carrying suspended solids, is returned to the beginning of the wastewater treatment process.

Removal of Plant Nutrients

When treatment standards require the removal of plant nutrients from the sewage, it is often done as a tertiary step. Phosphorus in wastewater is usually present in the form of organic compounds and phosphates that can easily be removed by chemical precipitation. This process, however, increases the volume and weight of sludge. Nitrogen, another important plant nutrient, is present in sewage in the form of ammonia and nitrates. Ammonia is toxic to fish, and it also exerts an oxygen demand in receiving waters as it is converted to nitrates. Nitrates, like phosphates, promote the growth of algae and the eutrophication of lakes. A method called nitrification-denitrification can be used to remove the nitrates. It is a two-step biological process in which ammonia nitrogen is first converted into nitrates by microorganisms. The nitrates are further metabolized by another species of bacteria, forming nitrogen gas that escapes into the air. This process requires the construction of more aeration and settling tanks and significantly increases the cost of treatment.

A physicochemical process called ammonia stripping may be used to remove ammonia from sewage. Chemicals are added to convert ammonium ions into ammonia gas. The sewage is then cascaded down through a tower, allowing the gas to come out of solution and escape into the air. Stripping is less expensive than nitrification-denitrification, but it does not work very efficiently in cold weather.

Land Treatment

In some locations, secondary effluent can be applied directly to the ground and a polished effluent obtained by natural processes as the wastewater flows over vegetation and percolates through the soil. There are three types of land treatment: slow-rate, rapid infiltration, and overland flow.

In the slow-rate, or irrigation, method, effluent is applied onto the land by ridge-and-furrow spreading (in ditches) or by sprinkler systems. Most of the water and nutrients are absorbed by the roots of growing vegetation. In the rapid infiltration method, the wastewater is stored in large ponds called recharge basins. Most of it percolates to the groundwater, and very little is absorbed by vegetation. For this method to work, soils must be highly permeable. In overland flow, wastewater is sprayed onto an inclined vegetated terrace and slowly flows to a collection ditch. Purification is achieved by physical, chemical, and biological processes, and the collected water is usually discharged into a nearby stream.

Land treatment of sewage can provide moisture and nutrients for the growth of vegetation, such as corn or grain for animal feed. It also can recharge, or replenish, groundwater aquifers. Land treatment, in effect, allows sewage to be recycled for beneficial use. Large land areas are required, however, and the feasibility of this kind of treatment may be limited further by soil texture and climate.

Clustered Wastewater Treatment Systems

In certain instances when it is not feasible to connect residences or units to public sewer systems, communities may opt for a clustered wastewater treatment system. Such facilities are smaller versions of centralized treatment plants and serve only a limited number of connections. The technologies used for clustered wastewater treatment may be the same as those used for centralized systems or for individual on-site systems, depending upon the specific applications and degree of treatment required. Upon treatment, effluent from clustered wastewater systems can be discharged via surface or subsurface disposal methods.

On-site Septic Tanks and Leaching Fields

In sparsely populated suburban or rural areas, it is usually not economical to build sewage collection systems and a centrally located treatment plant. Instead, a separate treatment and disposal system is provided for each home. On-site systems provide effective, low-cost, long-term solutions for wastewater disposal as long as they are properly designed, installed, and maintained. In the United States, about one-third of private homes make use of an on-site subsurface disposal system.

The most common type of on-site system includes a buried, watertight septic tank and a subsurface absorption field (also called a drain field or leaching field). The septic tank serves as a primary sedimentation and sludge storage chamber, removing most of the settleable and floating material from the influent wastewater. Although the sludge decomposes anaerobically, it eventually accumulates at the tank bottom and must be pumped out periodically (every two to four years). Floating solids and grease are trapped by a baffle at the tank outlet, and settled sewage flows out into the absorption field, through which it percolates downward into the ground. As it flows slowly through layers of soil, the settled wastewater is further treated and purified by both physical and biological processes before it reaches the water table.

An absorption field includes several perforated pipelines placed in long, shallow trenches filled with gravel. The pipes distribute the effluent over a sizable area as it seeps through the gravel and into the underlying layers of soil. If the disposal site is too small for a conventional leaching field, deeper seepage pits may be used instead of shallow trenches; seepage pits require less land area than leaching fields. Both leaching field trenches and seepage pits must be placed above seasonally high groundwater levels.

For subsurface on-site wastewater disposal to succeed, the permeability, or hydraulic conductivity, of the soil must be within an acceptable range. If it is too low, the effluent will not be able to flow effectively through the soil, and it may seep out onto the surface of the absorption field, thereby endangering public health. If permeability is too high, there may not be sufficient purification before the effluent reaches the water table, thereby contaminating the groundwater. The capacity of the ground to absorb settled wastewater depends largely on the texture of the soil (i.e., relative amounts of gravel, sand, silt, and clay). Permeability can be evaluated by direct observation of the soil in excavated test pits and also by conducting a percolation test, or "per test". The perc test measures the rate at which water seeps into the soil in small test holes dug on the disposal site. The measured perc rate can be used to determine the total required area of the absorption field or the number of seepage pits.

Where unfavourable site or soil conditions prohibit the use of both absorption fields and seepage pits, mound systems may be utilized for on-site sewage disposal. A mound is an absorption field built above the natural ground surface in order to provide suitable material for percolation and to separate the drain field from the water table. Septic tank effluent is intermittently pumped from a chamber and applied to the mound. Other alternative on-site disposal methods include use of intermittent sand filters or of small, prefabricated aerobic treatment units. Disinfection (usually by chlorination) of the effluent from these systems is required when the effluent is discharged into a nearby stream.

Wastewater Reuse

Wastewater can be a valuable resource in cities or towns where population is growing and water supplies are limited. In addition to easing the strain on limited freshwater supplies, the reuse of wastewater can improve the quality of streams and lakes by reducing the effluent discharges that they receive. Wastewater may be reclaimed and reused for crop and landscape irrigation, ground-water recharge, or recreational purposes. Reclamation for drinking is technically possible, but this reuse faces significant public resistance.

There are two types of wastewater reuse: direct and indirect. In direct reuse, treated wastewater is piped into some type of water system without first being diluted in a natural stream or lake or in groundwater. One example is the irrigation of a golf course with effluent from a municipal waste-water treatment plant. Indirect reuse involves the mixing of reclaimed wastewater with another body of water before reuse. In effect, any community that uses a surface water supply downstream from the treatment plant discharge pipe of another community is indirectly reusing wastewater. Indirect reuse is also accomplished by discharging reclaimed wastewater into a groundwater aqui-fer and later withdrawing the water for use. Discharge into an aquifer (called artificial recharge) is done by either deep-well injection or shallow surface spreading.

Quality and treatment requirements for reclaimed wastewater become more stringent as the chances for direct human contact and ingestion increase. The impurities that must be removed depend on the intended use of the water. For example, removal of phosphates or nitrates is not necessary if the intended use is landscape irrigation. If direct reuse as a potable supply is intended, tertiary treatment with multiple barriers against contaminants is required. This may include sec-ondary treatment followed by granular media filtration, ultraviolet radiation, granular activated carbon adsorption, reverse osmosis, air stripping, ozonation, and chlorination.

The use of gray-water recycling systems in new commercial buildings offers a method of saving water and reducing total sewage volumes. These systems filter and chlorinate drainage from tubs and sinks and reuse the water for nonpotable purposes (e.g., flushing toilets and urinals). Recycled water can be marked with a blue dye to ensure that it is not used for potable purposes.

Emerging Technologies

Experts in the wastewater treatment sector have been working to implement established technol-ogies and to improve environmental rules and regulations to meet water quality goals and human health protection. At the same time, the industry has also been transitioning to prepare for future challenges, such as climate change, changing populations, and aging infrastructure.

Improved Treatment Methods

Many older wastewater treatment facilities require upgrading because of increasingly strict wa-ter quality standards, but this is often difficult because of limited space for expansion. In order to allow improvement of treatment efficiencies without requiring more land area, new treatment methods have been developed. These include the membrane bioreactor process, the ballasted floc reactor, and the integrated fixed-film activated sludge (IFAS) process.

In the membrane bioreactor process, hollow-fibre microfiltration membrane modules are sub-merged in a single tank in which aeration, secondary clarification, and filtration can occur, thereby providing both secondary and tertiary treatment in a small land area.

In a ballasted floc reactor, the settling rate of suspended solids is increased by using sand and a polymer to help coagulate the suspended solids and form larger masses called flocs. The sand is separated from the sludge in a hydroclone, a relatively simple apparatus into which the water is introduced near the top of a cylinder at a tangent so that heavy materials such as sand are "spun" by centrifugal force toward the outside wall. The sand collects by gravity at the bottom of the hydroclone and is recycled back to the reactor.

Biological aerated filters use a basin with submerged media that serves as both a contact surface for biological treatment and a filter to separate solids from the wastewater. Fine-bubble aeration is applied to facilitate the process, and routine backwashing is used to clean the media. The land area required for a biological aerated filter is only about 15 percent of the area required for a conventional activated sludge system.

Automation

Advanced wastewater purification processes involve biological treatments that are sensitive to processing parameters and to the environment. To ensure stable and reliable operations of physical, chemical, and biological processes, treatment plants quite often need to implement sophisticated technologies involving complex instrumentation and process control systems. Use of online analytical instruments, programmable logic controllers (PLC), supervisory control and data acquisition (SCADA) systems, human machine interface (HMI), and various process control software allow for the automation and computerization of treatment processes with the provision for remote operations. Such innovations improve system operations significantly, thus minimizing supervision needs.

Environmental Considerations

Natural treatments, energy conservation, and carbon footprint reduction are some of the key considerations for communities facing energy and electricity challenges. Green technologies and the use of renewable energy sources, including solar and wind power, for wastewater treatment are evolving and will help minimize the environmental impacts of human activities. Ecological and economical natural wastewater treatment and disposal systems have already gained importance in many places, especially in smaller communities. These include constructed wetlands, lagoons, stabilization ponds, soil filters, drip irrigation, groundwater recharge, and other similar systems. The simplicity, cost-effectiveness, efficiency, and reliability of these systems have provided potential applications for such environmentally friendly technologies.

Given that wastewater is rich in nutrients and other chemicals, sewage treatment facilities have gained recognition as resource recovery facilities, overcoming their former reputation as mere pollution mitigation entities. Newer technologies and approaches have continued to improve the efficiency by which energy, nutrients, and other chemicals are recovered from treatment plants, helping create a sustainable market and becoming a revenue generation source for wastewater processing facilities.

Concepts such as nutrient trading have also emerged. The intention of such initiatives is to control and meet overall pollution load targets for a given watershed by trading nutrient reduction credits between point and non-point source dischargers. Such programs can help to minimize nutrient pollution effects as well as reduce financial burdens on societies for costly treatment plant upgrades.

Wastewater Treatment Systems

Oxidation Ditch

Oxidation ditch method, first developed in Netherlands, is a suitable method for the treatment of sewage in small communities. This is basically an aeration type of activated sludge process with a mechanical system of aeration. However, there is no primary sedimentation of sewage; consequently the problem of handling and treatment of primary sludge is eliminated.

Oxidation ditch consists of aeration units, namely ditch channels (2 or more) constructed side by side. The sizes of the ditch channels are variable-length 150-1000 m, width 1-5 m and depth 1-5 m. They are constructed with brick or stone masonry. A special type of rotors (cage rotors) are fitted into each ditch channel for continuously agitating and circulating the sludge, besides supplying O_2. The sludge is allowed to settle down in a separate sedimentation tank. The activated sludge is returned to ditch channels.

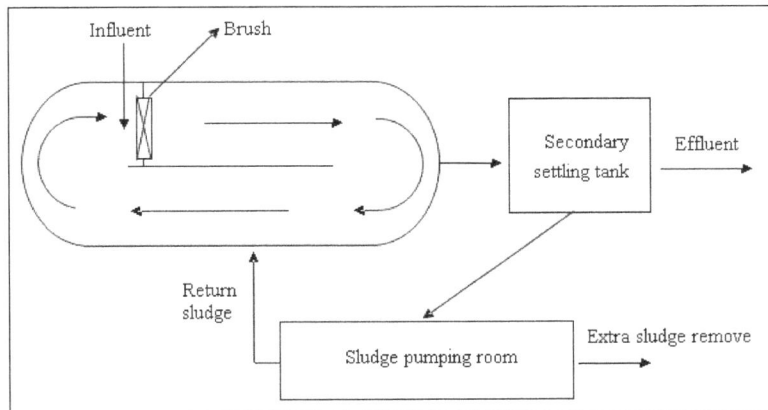

A Diagramic representation of oxidation ditch.

Instead of using a separate sedimentation tank, the sewage may be allowed to settle down in the ditch channels by stopping the rotors (usually during night). The supernatant from the ditch channels can be taken out. The excess sludge collected (in the sedimentation tank or at the bottom of ditch channels) can be stabilized.

Septic Tanks

Septic tanks are recommended for individual houses and for small communities and institutions with a contributing population around 300. Septic tanks work on the principle of anaerobic digestion. This occurs as the solids of the sewage settle at the bottom of the tank. Under anaerobic conditions, the biodegradable organic matter is converted to gases (CH_4, CO_2, H_2S etc.) and liquid compounds. This results in a drastic reduction in the volume of sludge.

A thick crust of scum formed on the surface of septic tank maintains anaerobic conditions. The effluent coming out of the septic tank contains some organic solids and pathogens. Therefore, disposal of the effluent from septic tank has to be dealt with very carefully. The septic tanks are dislodged and cleaned at regular intervals, usually once in 2-5 years (depending on the tank size and its use).

Construction and Operation of Septic Tanks

Septic tanks are usually constructed with bricks, or stone masonry. Thick-wall polythene and fibre glass tanks are also in use in recent years. Whatever may be the construction material used, the septic tank must be water-tight and must function efficiently. The size of the tank is variable, depending on the number of users. For a family of five members, a tank with a length of 1.5 m and a breadth of 0.75 m is recommended. For such a tank, the cleaning interval is 2-3 years.

A conventional septic tank has two compartments, the first compartment being twice the size of the second one. For effective sedimentation of solids, the tank should be designed to prevent the short-circuiting at the top and the bottom of the tank.

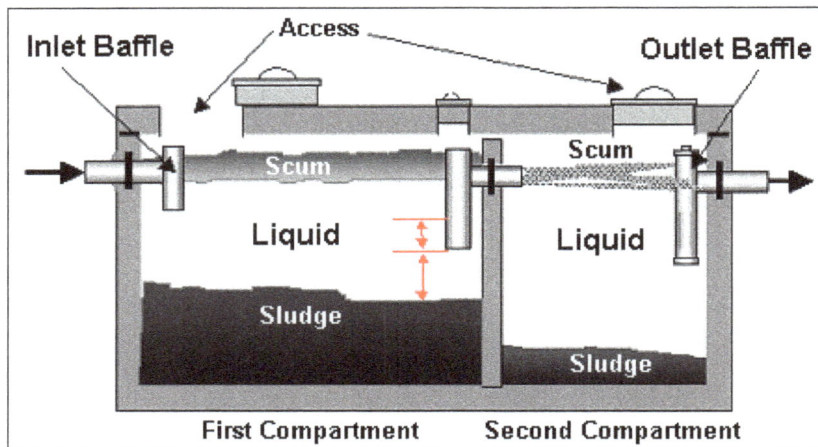

Schematic representation of a conventional two-compartment septic tank.

Further, the location of inlet and outlet should be such that the contents of the septic tank are not disturbed while the sewage enters or effluent leaves. Septic tank should be provided with a ventilation pipe, the top of which should be covered with a mosquito proof wire mesh.

As the sewage enters the septic tank, the solids settle to the bottom while grease and other light materials float on the surface, and form a scum. The bottom-settled organic material undergoes facultative and anaerobic decomposition to form more stable compounds and gases (CH_4, CO_2, and H_2S). In this way, there is a continuous reduction of solids of sewage entering the septic tank. However, sludge accumulates at the bottom of the tank. Desludging of septic tank has to be carried out periodically (once in 2-5 years).

Imhoff Tanks

Imhoff tank is an improved septic tank. It basically consists of a two storey tank in which the sedimentation (settling) occurs in the upper tank while the digestion of the settled solids takes place in the lower compartment.

As the sewage enters the sedimentation tank, the solids settle down to the bottom and the sewage flows into the digestion tank through slopes and slot of the sedimentation tank. Gas produced in the digestion process escapes through gas vent. The sedimentation tank is designed in such a way that the gases and the gas-buoyed sludge particles raising from the sludge layer do not enter into it. The sludge collection at the bottom can be withdrawn periodically.

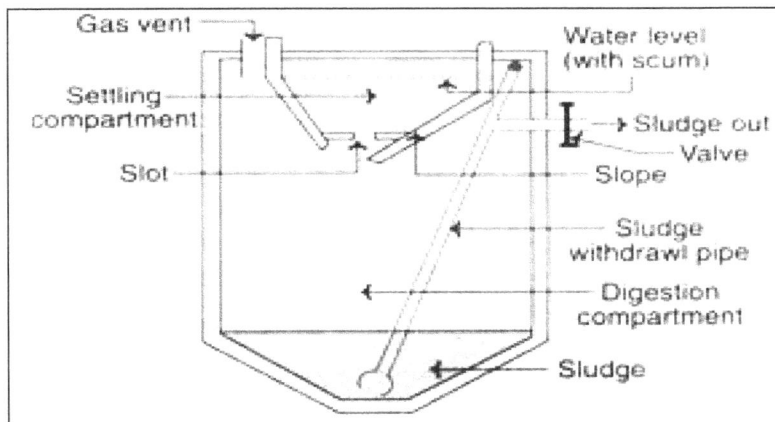

A diagrammatic view of imhoff tank.

Sedimentation Theory in Wastewater Treatment

Sedimentation (settling) is the separation of suspended particles that are heavier than water. The sedimentation of particles are based on the gravity force from the differences in density between particles and the fluid. Sedimentation is widely used in wastewater treatment systems. A successful sedimentation is crucial for the overall efficiency of the plant. Common examples include the removal of:

- Grit and particulate matter in the primary settling basin (settling tanks that receive raw wastewater prior to biological treatment are called primary tanks, forsedimentering).

- Sludge from the bioreactor (activated sludge process).

- Chemical flocs in the chemical step.

Often, the settler connected to the activated sludge process is the main bottle neck in the plant. The seemingly simple process has proven to be the weak link in many wastewater treatment plants.

The implementation of nitrogen removal in many Swedish plants emphasis the importance of the settler. The slow growth of nitrifying bacteria means that a high sludge age is necessary in the activated sludge process. For a give volume of the aeration basin, the sludge age may be increased by using a higher sludge concentration in the basin. However, by increasing the sludge concentration in the aeration basin, the capacity of the settler may be reached, the sludge blanket level will then increase which finally results in an uncontrolled sludge escape in the effluent water. Hence, there is a possible conflict between operation for good nutrient removal (high sludge age) and operation for good sludge sedimentation. Further, nitrogen removal in the activated sludge process gives also a risk for sludge rise in the secondary settler due to denitrification in the bottom of the settler. The sludge may rise due to flotation of solids when nitrogen gas is released.

Note also that the settler has two functions; clarification and sludge separation. That is to remove essentially all of the solids from suspension and to concentrate theses solids (eg for recycling to the aeration basin).

Depending on the particles concentration and the interaction between particles, four types of settling can occur:

- Discrete particle settling: The particles settle without interaction and occurs under low solids concentration. A typical occurrence of this type of settling is the removal of sand particles.

- Flocculent settling: This is defined as a condition where particles initially settle independently, but flocculate in the depth of the clarification unit. The velocity of settling particles are usually increasing as the particles aggregates. The mechanisms of flocculent settling are not well understood.

- Hindered settling: Inter-particle forces are sufficient to hinder the settling of neighboring particles. The particles tend to remain in a fixed positions with respect to each others. This type of settling is typical in the settler for the activated sludge process (secondary clarier).

- Compression settling: This occurs when the particle concentration is so high that so that particles at one level are mechanically influenced by particles on lower levels. The settling velocity then drastically reduces.

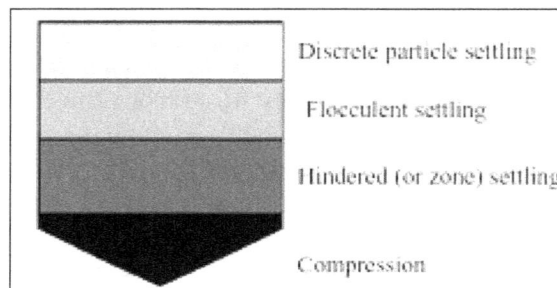
Settling phenomena in a clarifier.

Discrete Particle Settling

Consider the settling of a discrete particle. The sedimentation is obtained by the Newton and Stokes law.

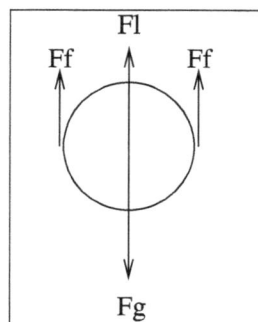
Forces on a discrete particle, F_g is the gravitational force, F_l isthe "lifting" force from the liquid, and F_f is the frictional force between the particle and the liquid.

Newton second law gives:

$$m\frac{dv}{dt} = F_g - F_l - F_f$$

where v is the velocity of the particle, and m is the mass. The gravity force F_g is given by:

$$F_g = mg = \rho_p V_p g$$

where ρ_p is the density of the particle and V_p is the volume.

The lifting force F_l is given by:

$$F_l = p_f = V_p g$$

where ρ_f is the density of the fluid.

The frictional drag force F_f depends on the particle velocity, fluid density, projected area, and a drag coecffient. The following empirical expression is used:

$$F_f = \frac{C_D A_p \rho_f v^2}{2}$$

where C_D is the drag coeffcient 1 and A_p is the projected area of the particle perpendicular to the velocity.

Inserting $F_g = mg = \rho_p V_p g$, $F_l = p_f = V_p g$ and $F_f = \frac{C_D A_p \rho_f v^2}{2}$ in $m\frac{dv}{dt} = F_g - F_l - F_f$ yields

$$m\frac{dv}{dt} = g(\rho_p - \rho_f)V_p - \frac{C_D A_p \rho_f v^2}{2}$$

In steady state $\left(\frac{dv}{dt} = 0\right)$ we have:

$$v = \sqrt{\frac{2g(\rho_p - \rho_f)V_p}{C_D A_p \rho_f}}$$

Stokes Law

For a spherical particle with diameter d, we have the volume $V_p = \frac{\pi d^3}{6}$ and a projected area $A_p = \frac{\pi d^2}{4}$ inserting this in $v = \sqrt{\frac{2g(\rho_p - \rho_f)V_p}{C_D A_p \rho_f}}$ gives:

$$v = \sqrt{\frac{4g(\rho_p - \rho_f)V_p}{3C_D \rho_f}}$$

For laminar flows it holds that:

$$C_D = \frac{24}{R_N}$$

where R_N is the Reynolds number.

$$R_N = \frac{vd\rho_f}{\mu}.$$

In $C_D = \frac{24}{R_N}$, the viscosity μ is introduced. This gives a measure of a fluids resistance to tangential or shear stress.

Inserting $C_D = \frac{24}{R_N}$ in $v = \sqrt{\frac{4g(\rho_p-\rho_f)V_p}{3C_D\rho_f}}$ using $R_N = \frac{vd\rho_f}{\mu}$ gives Stokes law:

$$v\frac{g(\rho_p-\rho_f)d^2}{18\mu}.$$

The Surface Loading Rate

The settling velocity of a particle can be used in the design of settling (sedimentation) basins. The key idea is to find a lower limit on the settling velocity for the particle to settle before it reach the outlet. Consider an ideal settling basins according to figure.

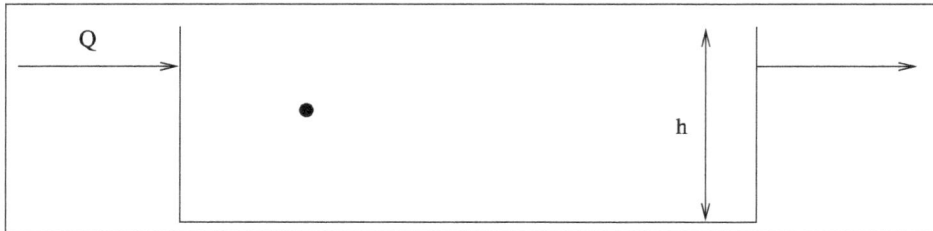

Settling in an ideal settling basin. The basin depth is h, the surface area is A.

The time a unit element is residencing in the (ideal) settling basin is given by:

$$T = \frac{V}{Q} = \frac{Ah}{Q}$$

where V is the basin volume, A is the surface of the basin, h is the basin depth, and Q is the flow rate. The minimum settling velocity for a particle (entering the basin at height h) to settle is thus given by:

$$v_{min} = h/T = \frac{Q}{A}.$$

The ratio $\frac{Q}{A}$ is the surface loading rate[2] and is one of the key parameters in the operation and design of settling basins. Note, however, that for normal settling basins the above relation only gives a crude approximation. Typically, only about 60% of the theoretical settling capacity $v_{min} = h/T = \frac{Q}{A}$

is achieved in practice. Different empirical relations exist to compensate for non ideal situations. Note that the basin depth h does not influence the (theoretical) minimum velocity. Common experience suggests that the basin depth should exceed 2-3m.

Hindered Settling

The Solid Flux Theory

Most models for hindered settling are based on the solid flux theory. Pioneering work, using solid flux concept for settler calculations, was done by Cloe and Clevenger and Kynch.

In general, the total flux (mass/(area time) of solids is obtained by:

$$J = Xv$$

where X is the solid (sludge) concentration[3] and v is the settling velocity which in general depends on X.

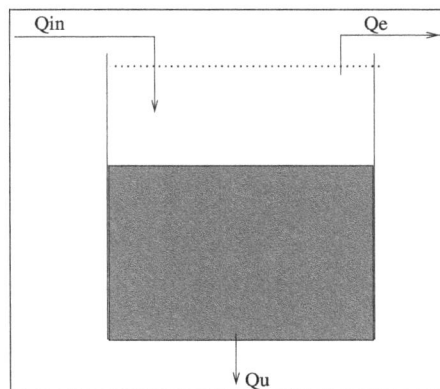

A settling basin. The basin has a cross sectional area A.
At the bottom of the basin, sludge is withdrawal at the rate Q_u.

The total flux of solids through a segment is:

$$J_t = J_g + J_u$$

where v_u is the velocity resulting from the removal of sludge at the bottom of the tank.. The solid flux theory states that:

$$J_g = v_g\left(X\right)X.$$

That is, the gravitational settling velocity only depends on the local concentration of solids. Hence, we can write the total flux as:

$$J_t = (v_g(X) + \frac{Q_u}{A})X.$$

Veslinds Formula

Several empirical relations have been suggested to describe the relation between the vg and X. A commonly used relation is the Veslind formula:

$$v_g\left(X\right) = v_o\ exp^{-nX}.$$

where v_o is the maximum settling velocity and n gives a measure on how fast the settling velocity decreases with increasing concentration of particles. In practice, these parameters can be found by multiple batch settling experiments where $\log v_g$ is measured for different sludge concentrations. Then, v_o and n can be found by a simple least squares fit to the data (linear regression). Also, several empirical relations exist where one tries to relate parameters like SV I (sludge volume index) to v_o and n.

Inserting the Veslind formula $v_g(X) = v_o \ exp^{-nX}$ in $J_t = (v_g(X) + \frac{Q_u}{A})X$ yield:

$$J_t = (v_o \ exp^{-nX} \frac{Q_u}{A})X$$

An illustration of the relation $J_t = (v_o \ exp^{-nX} \frac{Q_u}{A})X$ is given in figure.

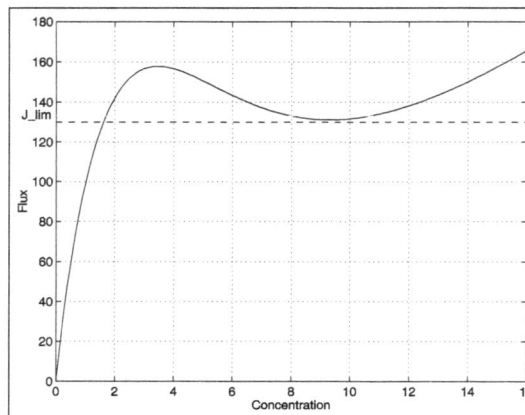

Total flux as a function of the concentration.

Notice that the flux curve has a local minima denoted J_{lim}. This flux is the maximum allowable flux loading if the settling is to be successful. If the influent flux to the settler is larger than J_{lim}, the sludge blanket will increase, resulting in solids (sludge) in the effluent.

For a general total flux model $J_t(X)$, the limiting flux can be obtained by solving $\frac{J_t(X)}{dX} = 0$. To find the minimum we have to check which of the extreme points that fulfill $\frac{J_t^2(X)}{dX^2} > 0$ These calculations may have to be solved numerically. Obviously, graphical solutions are also possible to use.

The limiting flux can be used for the design of the area of the settler. The ux at the level where the influent is located is:

$$J_{in} \frac{Q_{in}}{A} X_{in}$$

where X_{in} is the solid (sludge) concentration of the in uent water. We must require that $J_{in} < J_{lim}$ and hence:

$$A > \frac{Q_{in}}{J_{lim}} X_{in}$$

Note that J_{lim} in general depends on A.

Numerical Solution of the Settler

In order to model the settler during non-steady state conditions, the involved mass balance have to be solved. This is done by applying the conservation law (continuity equation) which states that the increase of mass per time unit equals the incoming flux minus the outgoing flux. This can be stated as a partial differential equation as:

$$\frac{\partial X}{\partial t} = -\frac{\partial J}{\partial z}$$

where z is the height coordinate.

The most popular approach to solve $\frac{\partial X}{\partial t} = -\frac{\partial J}{\partial z}$ numerically is to divide the settling basin in a number of horizontal slices (typically 10 to 100 slices). Each slice is regarded as well mixed. A further description of numerical methods is, however, outside the scope of this note.

Grit Removal

Municipal waste water contains a wide assortment of inorganic solids such as pebbles, sand, silt, egg shells, glass and metal fragments. Operations to remove these inorganic will also remove some of the larger, heavier organics such as bone chips, seeds etc. Together, these comprise the material known as grit in wastewater treatment systems.

Most of the substances in grit are abrasive in nature and will cause accelerated wear on pumps and sludge handling equipment with which it comes in contact Grit deposits in areas of low hydraulic shear in pipes, sumps and clarifiers may absorb grease and solidify. Also, these materials are not biodegradable and occupy valuable space in sludge digesters. It is, therefore, desirable to separate them from the organic suspended solids.

The latter should not be allowed to settle along with otherwise it gets entangled with the inorganic matter causing septicity of waste water and requiring unnecessary labour and expense for removal. A velocity of flow between 0.15 to 0.3 m/sec is practically considered sufficient for this purpose.

Grit removal facilities basically consist of an enlarged channel area where reduced flow velocities allow grit to settle out. Many configurations of grit tanks are available. At least two separate chambers should be provided, one to take care of low flow and the other for the high flow. A period of detention of 1 minute is commonly employed. Grit chambers are cleaned by hand, mechanically or hydraulically.

Hand cleaning is done only in the case of smaller plants, is less hygienic and odor free though somewhat easier for disposing off the removed material than in the case of mechanical cleaning. In hydraulic-cleaning, the deposited material is flushed out under fire-streams directed from a central point and removed through pipes in the side-walls or bottom of the chamber.

In larger treatment plants, the trend is towards aerated grit chamber. Turbulence created by the injection of compressed air keeps lighter organic material in suspension while the heavier grit falls to the bottom. Aerated grit chambers may serve another useful purpose.

If the sewage is anaerobic when it arrives at the plant, aeration serves to strip noxious gases from the liquid and to restore it immediately to an anaerobic condition, which allows for better treatment. When an aerated grit chamber is used for this purpose, the aeration period is usually extended from 15 to 20 minutes.

Grit particularly from channel-type grit chamber, may contain a sizeable fraction of biodegradable organics that must be removed by washing or must be disposed of quickly to avoid nuisance problems. Grit containing organics must either be placed in a sanitary landfill or incinerated, along with screenings, to a sterile ash for disposal.

Wastewater Chlorination

Disinfection of municipal wastewater is necessary for safe potable water supplies and for healthy rivers and streams. Microorganisms are present in large numbers in sewage treatment plant effluents and waterborne disease outbreaks have been associated with sewage-contaminated water supplies or recreational waters.

Chlorination is by far the most common method of wastewater disinfection and is used worldwide for the disinfection of pathogens before discharge into receiving streams, rivers or oceans. Chlorine is known to be effective in destroying a variety of bacteria, viruses and protozoa, including Salmonella, Shigella and Vibrio cholera.

Wastewater chlorination was initially applied in 1910 in Philadelphia, PA, and was soon implemented in many other cities in the United States based on this early success.

Today, wastewater chlorination is widely practiced to reduce microbial contamination and potential disease risks to exposed populations.

There is a water use cycle in which drinking water is treated, then consumed and discharged as wastewater. Following additional treatment, wastewater is discharged and may enter source waters used for drinking and recreation. Then the treatment-use-discharge process begins again, continuing the water use cycle.

Risks from Wastewater Contamination

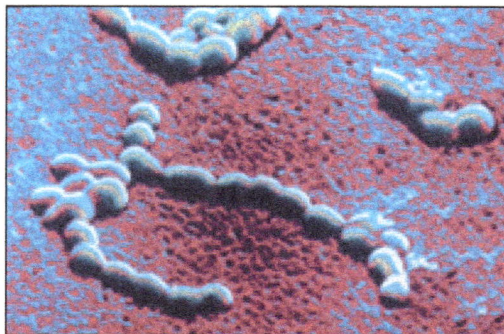

Wastewater can be discharged from a treatment plant into the environment wherehuman exposure may occur through the potable (drinking) water supply,recreation (swimming, snorkeling, etc.) or eating shellfish.

Pathogens commonly found in wastewater effluents are E. coli, Streptococcus, Salmonella, Shigella, mycobacteria, Pseudomonas aeroginosa, Giardia lamblia and enteroviruses. Tacnia, Ascaris and hookworm ova may be present in raw sewage. All of these microorganisms can make people sick.

The US Centers for Disease Control and Prevention has recorded a number of cases of shigellosis outbreaks caused by the consumption of freshwater shellfish harvested from waters contaminated by wastewater effluent. Natural decomposition processes would normally reduce these pathogens due to decay, predation and dilution. However, increasing human populations and discharge of effluent into receiving waters have limited the natural capability of self purification, making it necessary to disinfect the effluents before they are discharged.

Disinfection to remove or inactivate microorganisms is the most important step in wastewater treatment to prevent downstream users from contracting waterborne infectious diseases caused by microbes traditionally present in wastewaters. For example, wastewater disinfection helps prevent the accumulation of toxic microorganisms in fish, shellfish and other aquatic organisms. And, where sources of drinking water supplies may be contaminated by wastewater effluent, the importance of applying disinfection both upstream and downstream cannot be overstated.

The application of chlorine or any chemical disinfectant to wastewater results in the formation of by-products. The nature of biological treatment (prior to chlorination) and the presence of ammonia may have a substantial impact on the extent of by-product formation. Dechlorination of excess chlorine normally is performed prior to discharge to help prevent harm to aquatic life and also to reduce the formation and impact of disinfection by-products.

Chlorine: The Effective Disinfectant

Chlorination plays a key role in the wastewater treatment process by removing pathogens and other physical and chemical impurities. Chlorine's important benefits to wastewater treatment are listed below:

- Disinfection,
- Controlling odor and preventing septicity,
- Aiding scum and grease removal,
- Controlling activated sludge bulking,
- Controlling foaming and filter flies,
- Stabilizing waste activated sludge prior to disposal,
- Foul air scrubbing,
- Destroying cyanides and phenols,
- Ammonia removal.

Safe Handling and use

Chlorine can be used in wastewater disinfection as either elemental chlorine (gas) or as a chlorinated compound such as liquid sodium hypochlorite solution or solid calcium hypochlorite. Elemental chlorine is generally the most cost-effective option, but other factors must also be considered. In weakly buffered waters, for example, elemental chlorine depresses the pH, while the hypochlorites slightly raise the pH. The speed and efficacy of chlorine disinfection against pathogens may be affected by the pH of the water being treated.

Water Purification

Water purification is the process by which undesired chemical compounds, organic and inorganic materials, and biological contaminants are removed from water. That process also includes distillation (the conversion of a liquid into vapour to condense it back to liquid form) and deionization (ion removal through the extraction of dissolved salts). One major purpose of water purification is to provide clean drinking water. Water purification also meets the needs of medical, pharmacological, chemical, and industrial applications for clean and potable water. The purification procedure reduces the concentration of contaminants such as suspended particles, parasites, bacteria, algae, viruses, and fungi. Water purification takes place on scales from the large (e.g., for an entire city) to the small (e.g., for individual households).

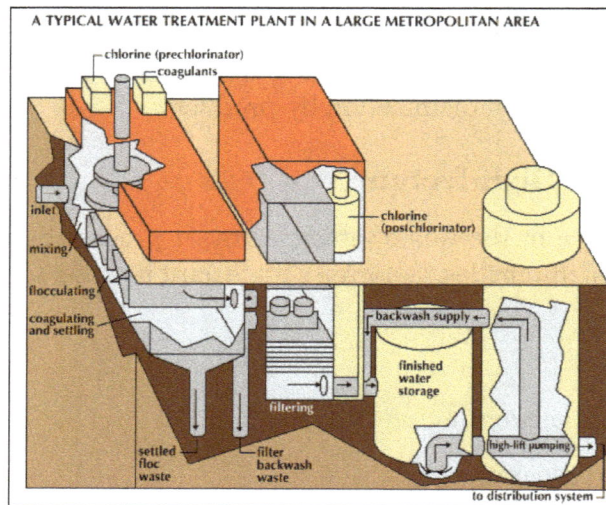

Water from inlets located in the water supply, such as a lake, is sent to be mixed, coagulated, and flocculated and is then sent to the waterworks for purification by filtering and chemical treatment. After being treated it is pumped into water mains for storage or distribution.

Most communities rely on natural bodies of water as intake sources for water purification and for day-to-day use. In general, these resources can be classified as groundwater or surface water and commonly include underground aquifers, creeks, streams, rivers, and lakes. With recent technological advancements, oceans and saltwater seas have also been used as alternative water sources for drinking and domestic use.

Determining Water Quality

Historical evidence suggests that water treatment was recognized and practiced by ancient civilizations.

In modern times, the quality to which water must be purified is typically set by government agencies. Whether set locally, nationally, or internationally, government standards typically set maximum concentrations of harmful contaminants that can be allowed in safe water. Since it is nearly impossible to examine water simply on the basis of appearance, multiple processes, such as physical, chemical, or biological analyses, have been developed to test contamination levels. Levels of organic and inorganic chemicals, such as chloride, copper, manganese, sulfates, and zinc, microbial pathogens, radioactive materials, and dissolved and suspended solids, as well as pH, odour, colour, and taste, are some of the common parameters analyzed to assess water quality and contamination levels.

Regular household methods such as boiling water or using an activated-carbon filter can remove some water contaminants. Although those methods are popular because they can be used widely and inexpensively, they often do not remove more dangerous contaminants. For example, natural spring water from artesian wells was historically considered clean for all practical purposes, but it came under scrutiny during the first decade of the 21st century because of worries over pesticides, fertilizers, and other chemicals from the surface entering wells. As a result, artesian wells were subjected to treatment and batteries of tests, including tests for the parasite Cryptosporidium.

Not all people have access to safe drinking water. According to a 2017 report by the United Nations (UN) World Health Organization (WHO), 2.1 billion people lack access to a safe and reliable drinking water supply at home. Eighty-eight percent of the four billion annual cases of diarrhea reported worldwide have been attributed to a lack of sanitary drinking water. Each year approximately 525,000 children under age five die from diarrhea, the second leading cause of death, and 1.7 million are sickened by diarrheal diseases caused by unsafe water, coupled with inadequate sanitation and hygiene.

Process

Most water used in industrialized countries is treated at water treatment plants. Although the methods those plants use in pretreatment depend on their size and the severity of the contamination, those practices have been standardized to ensure general compliance with national and international regulations. The majority of water is purified after it has been pumped from its natural source or directed via pipelines into holding tanks. After the water has been transported to a central location, the process of purification begins.

Pretreatment

In pretreatment, biological contaminants, chemicals, and other materials are removed from water. The first step in that process is screening, which removes large debris such as sticks and trash from the water to be treated. Screening is generally used when purifying surface water such as that from lakes and rivers. Surface water presents a greater risk of having been polluted with large amounts of contaminants. Pretreatment may include the addition of chemicals to control the growth of bacteria in pipes and tanks (prechlorination) and a stage that incorporates sand filtration, which helps suspended solids settle to the bottom of a storage tank.

Preconditioning, in which water with high mineral content (hard water) is treated with sodium carbonate (soda ash), is also part of the pretreatment process. During that step, sodium carbonate is added to the water to force out calcium carbonate, which is one of the main components in shells of marine life and is an active ingredient in agricultural lime. Preconditioning ensures that hard water, which leaves mineral deposits behind that can clog pipes, is altered to achieve the same consistency as soft water.

Prechlorination, which is often the final step of pretreatment and a standard practice in many parts of the world, has been questioned by scientists. During the prechlorination process, chlorine is applied to raw water that may contain high concentrations of natural organic matter. This organic matter reacts with chlorine during the disinfection process and can result in the formation of disinfection by-products (DBPs), such as trihalomethanes, haloacetic acids, chlorite, and bromate. Exposure to DBPs in drinking water can lead to health issues. Worries stem from the practice's possible association with stomach and bladder cancer and the hazards of releasing chlorine into the environment.

Other Purification Steps

After pretreatment, chemical treatment and refinement can occur. That process includes coagulation, a step in which chemicals are added that cause small particles suspended in the water to clump together. Flocculation follows, which mixes the water with large paddles so that coagulated particles can be brought together into larger clumps (or "floc") that slowly settle on the bottom of the tank or basin.

After the majority of the suspended particles have settled, water exits the flocculation basin and then enters a sedimentation basin. Sedimentation basins move treated waters along through the purification process while allowing remaining particles to settle. Sludge forms that appear on the floor of the tank are removed and treated. From that basin, water is moved to the next step, filtration, which removes the remaining suspended particles and unsettled floc in addition to many microorganisms and algae.

Disinfection is the final step in water purification. During that step, harmful microbes, such as bacteria, viruses, and protozoa, are killed through the addition of disinfectant chemicals. Disinfection usually involves a form of chlorine, especially chloramines or chlorine dioxide. Chlorine is a toxic gas, resulting in some danger from release associated with its use. To avoid those risks, some water treatment plants use ozone, ultraviolet radiation, or hydrogen peroxide disinfection instead of chlorine. Other purification methodologies include ultrafiltration for specific dissolved substances, ion exchange to remove metal ions, and fluoridation to prevent tooth decay.

In certain areas of the world that do not have access to water treatment plants, alternative methods of purification must be used. Those methods include boiling, granular activated-carbon filtering, distillation, reverse osmosis, and direct contact membrane distillation.

Industrial Water Purification

In addition to drinking and domestic uses, industries also consume significant amounts of water. Chemical, petroleum, food processing, and textile industries, for example, require water for manufacturing, processing, heating, cooling, washing, rinsing, and other applications. Such industrial systems require treated water, and the lack of appropriate purification can lead to issues such as scaling, corrosion, deposition, bacterial growth within piping or processing equipment, and poor product quality. In addition to conventional water treatment processes, industrial water purification may also involve specialized techniques such as electrodeionization, ion exchange, membrane systems, ozone treatment, evaporation, and ultraviolet irradiation. Technologies selection depends upon the raw water quality and the intended industrial use.

Saline Water Purification

The vast majority of communities rely on freshwater resources for drinking and domestic water supplies. However, with shrinking freshwater reserves and rising water demands complicated by natural factors such as droughts, floods, and climate change impacts, several countries have begun to utilize oceans and inland seas as alternative water sources. Desalination technologies that remove salts and minerals from seawater are emerging to produce potable water suitable for drinking and domestic purposes. Reverse osmosis, vacuum distillation, multistage flash distillation, freeze-thaw, and electrodialysis are gaining importance for saltwater purification. Such processes usually involve higher energy consumption and are comparatively more expensive than conventional freshwater treatment processes. Numerous efforts are under way to make desalination methods cost-effective and economically viable.

System Configurations and Improvements

The size and capacity of water treatment systems vary widely, ranging from simple household units to small facilities that serve manufacturing industries to large-scale centralized water treatment plants dedicated to cities and towns. Selection of specific treatment processes depends upon factors such as intake water quality, degree of purification required, intended water use, flow capacity requirements, government regulations, available capital, and the operations and maintenance costs involved. Treated water is distributed to consumers via water distribution systems involving pipes, pumps, booster stations, storage tanks, and associated appurtenances.

Water purification plant.

In an effort to meet stringent environmental regulations and to satisfy the rising water demands of growing populations, many water treatment plants have employed smart technologies to increase operations reliability. Water sustainability improvements, which can increase the energy efficiency of a plant and reduce its carbon footprint, often include the optimization of chemical use, a minimization of waste generation, and the use of solar or wind energy. Additionally, with the advancement of sophisticated technologies, water treatment processes have incorporated complex instrumentation and process control systems. Use of online analytical instruments, supervisory control and data acquisition (SCADA) systems, and dedicated software have resulted in automation and computerization of treatment processes with the provision for remote operations. Such innovations can improve system operations significantly to achieve consistent water quality with minimal supervision, especially in larger system configurations.

Wastewater Chemical Treatment Processes

Chemicals are used during wastewater treatment in an array of processes to expedite disinfection. These chemical processes, which induce chemical reactions, are called chemical unit processes and are used alongside biological and physical cleaning processes to achieve various water standards.

Specialized chemicals such as chlorine, hydrogen peroxide, sodium chlorite, and sodium hypochlorite (bleach) act as agents that disinfect, sanitize, and assist in the purification of wastewater at treatment facilities.

There are several distinct chemical unit processes, including chemical coagulation, chemical precipitation, chemical oxidation, and advanced oxidation, ion exchange, and chemical neutralization and stabilization, which can be applied to wastewater during cleaning.

Chemical Precipitation

Chemical precipitation is the most common method for removing dissolved metals from wastewater solution containing toxic metals. To convert the dissolved metals into solid particle form, a precipitation reagent is added to the mixture. A chemical reaction, triggered by the reagent, causes the dissolved metals to form solid particles. Filtration can then be used to remove the particles from the mixture. How well the process works is dependent upon the kind of metal present, the concentration of the metal, and the kind of reagent used. In hydroxide precipitation, a commonly used chemical precipitation process, calcium or sodium hydroxide is used as the reagent to create solid metal hydroxides. However, it can be difficult to create hydroxides from dissolved metal particles in wastewater because many wastewater solutions contain mixed metals.

Chemical Coagulation

This chemical process involves destabilizing wastewater particles so that they aggregate during chemical flocculation. Fine solid particles dispersed in wastewater carry negative electric surface charges (in their normal stable state), which prevent them from forming larger groups and settling. Chemical coagulation destabilizes these particles by introducing positively charged coagulants that then reduce the negative particles' charge. Once the charge is reduced, the particles freely form larger groups. Next, an anionic flocculant is introduced to the mixture. Because the flocculant reacts against the positively charged mixture, it either neutralizes the particle groups or creates bridges between them to bind the particles into larger groups. After larger particle groups are formed, sedimentation can be used to remove the particles from the mixture.

Chemical Oxidation and Advanced Oxidation

With the introduction of an oxidizing agent during chemical oxidation, electrons move from the oxidant to the pollutants in wastewater. The pollutants then undergo structural modification, becoming less destructive compounds. Alkaline chlorination uses chlorine as an oxidant against cyanide. However, alkaline chlorination as a chemical oxidation process can lead to the creation of toxic chlorinated compounds, and additional steps may be required. Advanced oxidation can help remove any organic compounds that are produced as a byproduct of chemical oxidation, through processes such as steam stripping, air stripping, or activated carbon adsorption.

Ion Exchange

When water is too hard, it is difficult to use to clean and often leaves a grey residue. (This is why clothing washed in hard water often retains a dingy tint.) An ion exchange process, similar to the reverse osmosis process, can be used to soften the water. Calcium and magnesium are common ions that lead to water hardness. To soften the water, positively charged sodium ions are introduced in the form of dissolved sodium chloride salt or brine. Hard calcium and magnesium ions exchange places with sodium ions, and free sodium ions are simply released in the water. However, after softening a large amount of water, the softening solution may fill with excess calcium and magnesium ions, requiring the solution to be recharged with sodium ions.

Chemical Stabilization

This chemical wastewater treatment process works in a similar fashion as chemical oxidation. Sludge is treated with a large amount of a given oxidant, such as chlorine. The introduction of the oxidant slows down the rate of biological growth within the sludge and also helps deodorize the mixture. The water is then removed from the sludge. Hydrogen peroxide can also be used as an oxidant and may be a more cost-effective choice.

Physical Water Treatment Methods

Physical water treatment typically consists of filtration techniques that involve the use of screens, sand filtration or cross flow filtration membranes.

- Screens: Typically used as a pretreatment method to remove larger suspended material.

- Sand and/or Multi Media Filtration: Frequently used to filter suspended solids. Smaller suspended solids and dissolved solids are often able to pass through these filters, requiring secondary filtration.

- Membrane Filtration: Utilizes barrier (microfiltration, ultrafiltration) or semipermeable (nano or reverse osmosis) membranes to remove suspended solids and total dissolved solids, respectively.

Greensand Filtration

Glauconite is a mineral commonly referred to as green sand and is used in greensand filtration. It is an effective filtration medium for the removal of dissolved iron, hydrogen sulphide, and manganese from water. Glauconite is coated with manganese oxide, which causes soluble iron, manganese and hydrogen sulfide gas to bond with oxygen. Bonding with oxygen causes the previously dissolved elements to precipitate and become embedded in the greensand filter.

Multi Media Filtration (MMF)

Multimedia filtration is a modern physical water treatment technique that uses at least three different layers of filtration media, typically anthracite, sand and garnet, to filter water. This filter arrangement allows for larger particulates to be trapped at the top of the filter while smaller particulates are trapped deeper in the media. Suspended solids, including: clay, algae, silt, rust, and other organic matter are removed as the water passes through each layer of media. This filtration method is capable of removing particles from 10 to 25 microns in size. Multi media filtration does not remove viruses, bacteria or smaller protozoans.

Microfiltration

Unlike greensand and multimedia filters, microfiltration uses a barrier membrane to filter very small suspended solids from water. Microfiltration membranes are typically capable of removing contaminants ranging from 0.1 to 10 microns in size. This form of physical water treatment is ideal for removing suspended solids, algae and protozoans from water but does not generally remove bacteria and viruses. Microfiltration does not remove dissolved contaminants from water.

Ultrafiltration

Ultrafiltration is a physical water filtration process that utilizes pressure to separate solids from water through a barrier membrane. This filtration process is capable of removing suspended solids, bacteria and certain viruses ranging from 0.005 to 0.01 micron in size, and is sometimes used as a pretreatment method upstream of reverse osmosis. Ultrafiltration cannot remove dissolved solids.

Nanofiltration

Nanofiltration works similar to ultrafiltration, but utilizes a semipermeable membrane with an even smaller pore size. Nanofilters are capable of removing bacteria, viruses and divalent and multivalent ions (e.g. calcium, magnesium). It functions as a barrier membrane capable of removing particles ranging from 0.005 to 0.001 micron in size, and also acts as a semi-permeable membrane capable of removing ions. Due to its ability to remove divalent ions such as calcium it is sometimes referred to as the "softening membrane". Learn more about nanofiltration for industrial water treatment.

Reverse Osmosis

Reverse osmosis is one of the most common physical water treatment methods employed in industrial water treatment. Reverse osmosis, also known as RO, filters contaminants out of water using applied pressure to force water through a semipermeable membrane. RO is capable of removing impurities such as dissolved ions (e.g., sodium), bacteria, viruses, and other contaminants ranging from 0.005 to 0.0001 micron in size.

The most effective water treatment systems make use of a combination of biological, chemical and physical water treatment methods, as well as appropriate pretreatment and post-treatment methods to produce water that is free from unwanted contaminants.

Activated Sludge Process

Activated sludge (AS) is a process dealing with the treatment of sewage and industrial wastewaters and developed around 1912-1914. There is a large varity of design, however, in principle all AS consist of three main components: an aeration tank, which serves as bio reactor; a settling tank ("final clarifier") for seperation of AS solids and treated waste water; a return activated sludge (RAS) equipment to transfer settled AS from the clarifier to the influent of the aeration tank. Atmospheric air or in rare cases pure oxygen is introduced to a mixture of primary treated or screened sewage (or industrial wastewater) combined with organisms to develop a biological floc ("Activated Sludge" AS). The mixture of raw sewage (or industrial wastewater) and biological mass is commonly known as Mixed Liquor. Typically, dry solids concentrations of mixed liquor (MLSS) range from 3 to 6 g/L. With all activated sludge plants, the concentration of biodegradable components present in the influent is reduced due to biological (and sometimes chemical) processes in the aeration tank. The removal efficiency is controlled by different boundary conditions, e.g. the hydraulic residence time (HRT) in the aeration tank, which is defined by aeration tank volume divided by the flow rate. Other factors are: Influent load (BOD5, COD, Nitrogen) in relation to the AS solids present in the aeration tank (Food:Microorganism Ratio, F:M Ratio), oxygen supply, temperature, etc. At the effluent of the aeration tank, mixed liquor is discharged into settling tanks and the supernatant (treated waste water) is run off to be discharged to a natural water or undergo further treatment before discharge. The settled AS is returned to the head of the aeration tank (RAS) to re-seed the new sewage (or industrial wastewater) entering the tank and to ensure the desired MLSS concentration in the aeration tank. Due to biological growth (and solids present in the raw waste water which are only partly degraded), excess sludge eventually accumulates beyond the desired MLSS concentration in the aeration tank. This amount of solid (called Waste Activated Sludge WAS) is removed from the treatment process to keep the ratio of biomass to food supplied (sewage or wastewater) in balance and the F:M ratio in a defined range. WAS is stored away from the main treatment process in storage tanks and is further treated by digestion, either under anaerobic or aerobic conditions prior to disposal.

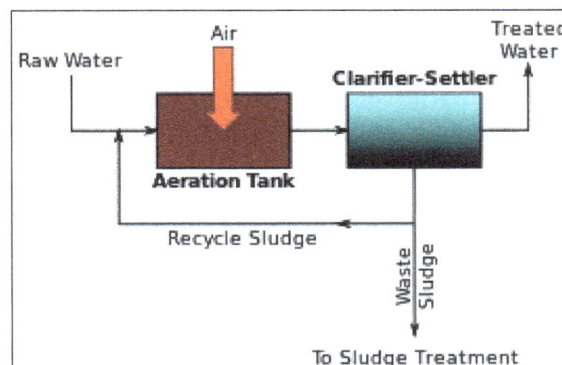

The diagram of activated sludge process.

Many sewage treatment plants use axial flow pumps to transfer nitrified mixed liquor from the aeration zone to the anoxic zone for de-nitrification. These pumps are often referred to as Internal Mixed Liquor Recycle pumps (IMLR pumps). The raw sewage, the RAS, and the nitrified mixed liquor are mixed by submersible mixers in the anoxic zones in order to achieve de-nitrification.

Purpose

- In a sewage (or industrial wastewater) treatment plant, the activated sludge process can be used for one or several of the following purposes.

- Oxidizing carbonaceous matter: Biological matter.

- Oxidizing nitrogeneous matter: Mainly ammonium and nitrogen in biological materials.

- Removing phosphate.

- Driving off entrained gases carbon dioxide, ammonia, nitrogen, etc.

- Generating a biological floc that is easy to settle.

- Generating a liquor low in dissolved or suspended material.

Activated Sludge Process Variables

The main variables of activated sludge process are the mixing regime, loading rate, and the flow scheme.

Mixing Regime

Generally two types of mixing regimes are of major interest in activated sludge process: plug flow and complete mixing. In the first one, the regime is characterized by orderly flow of mixed liquor through the aeration tank with no element of mixed liquor overtaking or mixing with any other element. There may be lateral mixing of mixed liquor but there must be no mixing along the path of flow.

In complete mixing, the contents of aeration tank are well stirred and uniform throughout. Thus, at steady state, the effluent from the aeration tank has the same composition as the aeration tank contents.

The type of mixing regime is very important as it affects (1) oxygen transfer requirements in the aeration tank, (2) susceptibility of biomass to shock loads, (3) local environmental conditions in the aeration tank, and (4) the kinetics governing the treatment process.

Flow Scheme

The flow scheme involves:

- The pattern of sewage addition.

- The pattern of sludge return to the aeration tank.

- The pattern of aeration.

Sewage addition may be at a single point at the inlet end or it may be at several points along the aeration tank. The sludge return may be directly from the settling tank to the aeration tank or through a sludge reaeration tank. Aeration may be at a uniform rate or it may be varied from the head of the aeration tank to its end.

Types of Plants

There are a variety of types of activated sludge plants. These include:

Package Plants

There are a wide range of other types of plants, often serving small communities or industrial plants that may use hybrid treatment processes often involving the use of aerobic sludge to treat the incoming sewage. In such plants the primary settlement stage of treatment may be omitted. In these plants, a biotic floc is created which provides the required substrate.

Package plants are commonly variants of extended aeration, to promote the 'fit & forget' approach required for small communities without dedicated operational staff. There are various standards to assist with their design.

Oxidation Ditch

In some areas, where more land is available, sewage is treated in large round or oval ditches with one or more horizontal aerators typically called brush or disc aerators which drive the mixed liquor around the ditch and provide aeration. These are oxidation ditches, often referred to by manufacturer's trade names such as Pasveer, Orbal, or Carrousel. They have the advantage that they are relatively easy to maintain and are resilient to shock loads that often occur in smaller communities (i.e at breakfast time and in the evening).

Oxidation ditches are installed commonly as 'fit & forget' technology, with typical design parameters of a hydraulic retention time of 24-48 hours, and a sludge age of 12-20 days. This compares with nitrifying activated sludge plants having a retention time of 8 hours, and a sludge age of 8-12 days.

Deep Shaft

Where land is in short supply sewage may be treated by injection of oxygen into a pressured return sludge stream which is injected into the base of a deep columnar tank buried in the ground. Such shafts may be up to 100 m deep and are filled with sewage liquor. As the sewage rises the oxygen forced into solution by the pressure at the base of the shaft breaks out as molecular oxygen providing a highly efficient source of oxygen for the activated sludge biota. The rising oxygen and injected return sludgeprovide the physical mechanism for mixing of the sewage and sludge. Mixed sludge and sewage is decanted at the surface and separated into supernatant and sludgecomponents. The efficiency of deep shaft treatment can be high.

Surface aerators are commonly quoted as having an aeration efficiency of 0.5-1.5 kg O2/kWh, diffused aeration as 1.5-2.5 kg O2/KWh. Deep Shaft claims 5-8 kg O2/kWh.

However, the costs of construction are high. Deep Shaft has seen greatest uptake in Japan, because of the land area issues. Deep Shaft was developed by ICI, as a spin-off from their Pruteen process.

In the UK it is found at three sites: Tilbury, Anglian water, treating a wastewater with a high industrial contribution; Southport, United Utilities, because of land space issues; and Billingham, ICI, again treating industrial effluent, and built (after the Tilbury shafts) by ICI to help the agent sell more.

Deep Shaft is a patented, licensed, process. The licensee has changed several times and, currently, it is Aker Kvaerner Engineering Services.

Surface-aerated Basins

A Typical Surface-Aerated Basing (using motor-driven floating aerators).

Most biological oxidation processes for treating industrial wastewaters have in common the use of oxygen (or air) and microbial action. Surface-aerated basins achieve 80 to 90% removal of BOD with retention times of 1 to 10 days. The basins may range in depth from 1.5 to 5.0 m and utilize motor-driven aerators floating on the surface of the wastewater.

In an aerated basin system, the aerators provide two functions: they transfer air into the basins required by the biological oxidation reactions, and they provide the mixing required for dispersing the air and for contacting the reactants (that is, oxygen, wastewater and microbes). Typically, the floating surface aerators are rated to deliver the amount of air equivalent to 1.8 to 2.7 kg O2/kWh. However, they do not provide as good mixing as is normally achieved in activated sludge systems and therefore aerated basins do not achieve the same performance level as activated sludge units.

Biological oxidation processes are sensitive to temperature and, between 0 °C and 40 °C, the rate of biological reactions increase with temperature. Most surface aerated vessels operate at between 4 °C and 32 °C.

Aeration Methods

Diffused Aeration

Sewage liquor is run into deep tanks with diffuser blocks attached to the floor. These are like the diffuser blocks used in tropical fish tanks but on a much larger scale. Air is pumped through the blocks and the curtain of bubbles formed both oxygenates the liquor and also provide the necessary stirring action. Where capacity is limited or the sewage is unusually strong or difficult to treat, oxygen may be used instead of air. Typically, the air is generated by some type of blower or compressor.

Surface Aerators

Vertically mounted tubes of up to 1 m diameter extending from just above the base of a deep concrete tank to just below the surface of the sewage liquor. A typical shaft might be 10 m high. At the surface end the tube is formed into a cone with helical vanes attached to the inner surface. When the tube is rotated, the vanes spin liquor up and out of the cones drawing new sewage liquor from the base of the tank. In many works each cone is located in a separate cell that can be isolated from the remaining cells if required for maintenance. Some works may have two cones to a cell and some large works may have 4 cones per cell.

Methods of Sewage Treatment

There are different types of sewage systems which can be described as on-site systems and sewage or effluent systems.

An on-site system is one which treats the sewage in a septic tank so that most of the sewage becomes effluent and is disposed of in an area close to the house or buildings. An example of an on-site disposal system consists of a septic tank and leach drains.

A sewage or wastewater system disposes of the effluent from a community at a central place usually called a sewage lagoon or effluent pond. The sewage can be treated:

- In a septic tank at each building.

- Just before the lagoon in a large septic tank or macerator system.

- In the lagoon itself.

On-site Disposal Systems

All the liquid waste from the toilet, bathroom, laundry and sink goes into pipes which carry it to a septic tank. The effluent from the tank is then disposed of through effluent disposal drains often referred to as leach or French drains. Both of these methods of disposing of liquid waste are on-site disposal systems. They must be installed and maintained properly.

In these systems, the effluent is soaked into the surrounding soil. Some soils don't allow good soakage such as clay or similar soils; if there are any problems with this disposal system a local government EHO should be consulted to talk about the problem.

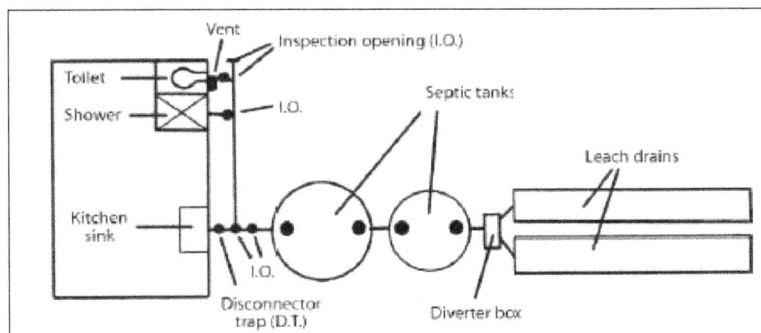

Plan view (top) of an on-site sewage disposal system.

On-site disposal systems cannot be installed in all situations. For example, they cannot be installed:

- In areas that flood regularly,

- In areas that have a high water table (that is, where the underground water is close to the surface),

- Where the amount of wastewater to be disposed of is large,

- Near to drinking water supplies.

Effluent (Wastewater) Disposal System

In this method the effluent from the community is carried by large pipes to the lagoon. These pipes serve all the houses and other buildings in the community. The sewage may be either be treated in septic tanks at the houses or buildings or at the lagoon. There are no leach or French drains.

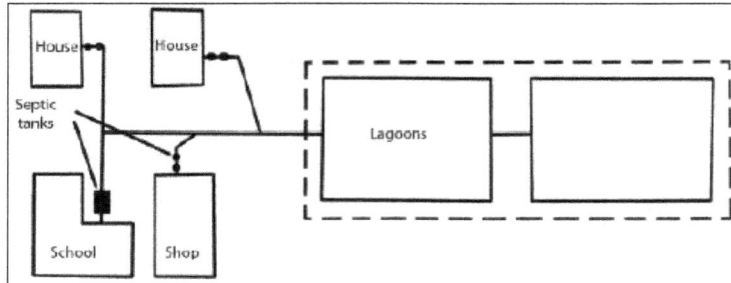

Plan view of a wastewater disposal system.

Full Sewage System

All the sewage from the toilet, shower, laundry and other areas enters waste and sewer pipes directly and is pumped to a lagoon.

- There are three types of full sewage system:

- The sewage enters the lagoon without treatment.

- The sewage goes through a series of cutting blades which help break up the solid matter before it enters the lagoon. These blades are called macerators.

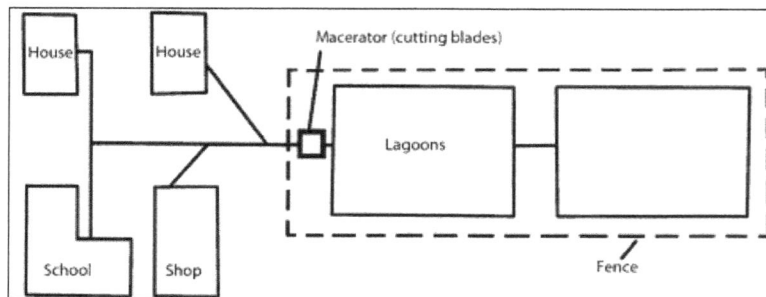

Plan view of full sewage system and macerators.

- The sewage may be treated in a large septic tank just before it enters the lagoon.

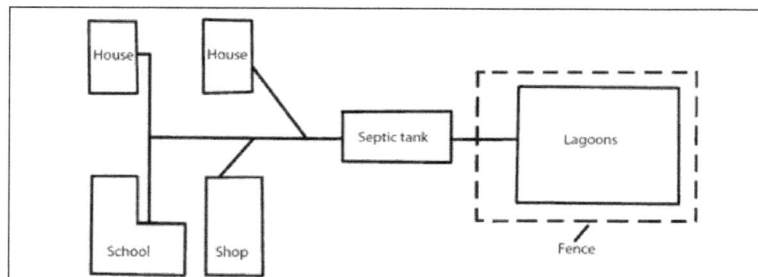

Plan view of a full sewage system with a large septic tank.

References

- Wastewater-treatment: britannica.com, Retrieved 03 February, 2019

- Waste-water-treatment-systems-with-diagram, waste-management-waste-water-treatment – 11003: biology-discussion.com, Retrieved 16 June, 2019

- Grit-removal-as-primary-treatment-of-waste-water, waste-management – 28306: yourarticlelibrary.com, Retrieved 08 January, 2019

- Wastewater-Chlorination: chlorine.americanchemistry.com, Retrieved 02 February, 2019

- Water-purification: britannica.com, Retrieved 19 June, 2019

- Wastewater-chemical-treatment: thomasnet.com, Retrieved 05 March, 2019

- Activated-sludge-process: iwapublishing.com, Retrieved 18 August, 2019

Chapter 4

Biological Wastewater Treatment

Biological wastewater treatment is the secondary treatment process which is used to remove sediments, oils, etc. which remain after primary treatment. These are divided into anaerobic and aerobic processes. This chapter closely examines the aspects associated with biological wastewater treatment to provide an extensive understanding of the subject.

Biological wastewater treatment harnesses the action of bacteria and other microorganisms to clean water.

Biological wastewater treatment is a process that seems simple on the surface since it uses natural processes to help with the decomposition of organic substances, but in fact, it's a complex, not completely understood process at the intersection of biology and biochemistry.

Biological treatments rely on bacteria, nematodes, or other small organisms to break down organic wastes using normal cellular processes. Wastewater typically contains a buffet of organic matter, such as garbage, wastes, and partially digested foods. It may also contain pathogenic organisms, heavy metals, and toxins.

The goal of biological wastewater treatment is to create a system in which the results of decomposition are easily collected for proper disposal. Scientists have been able to control and refine both aerobic and anaerobic biological processes to achieve the optimal removal of organic substances from wastewater.

Effective and Economical Treatment

Biological treatment is used worldwide because it's effective and more economical than many mechanical or chemical processes.

Biological wastewater treatment is often a secondary treatment process, used to remove any material remaining after primary treatment. In the primary water treatment process, sediments or substances such as oil are removed from the wastewater.

The biological processes used to treat wastewater include subsurface applications, such as septic or aerobic tank disposal systems; many types of aeration, including surface and spray aeration; activated sludge processes; ponds and lagoons; trickling filters; and anaerobic digestion. Constructed wetlands and various types of filtration are also considered biological treatment processes.

These processes are usually divided into anaerobic and aerobic processes. "Aerobic" refers to a process in which oxygen is present, while "anaerobic" describes a biological process in which oxygen is absent.

Aerobic Wastewater Treatment

Aerobic wastewater treatment processes include treatments such as activated sludge process, oxidation ditches, trickling filters, lagoon-based treatments, and aerobic digestion. Diffused aeration systems may be used to maximize oxygen transfer and minimize odors as the wastewater is treated. Aeration provides oxygen to the helpful bacteria and other organisms as they decompose organic substances in the wastewater.

A time-honored example of an aerobic treatment method is the activated sludge process. This is a proven biological wastewater treatment widely used for the secondary treatment of both domestic and industrial wastewater. It is well suited for treating waste streams high in organic or biodegradable content and is often used to treat municipal sewage, wastewater generated by pulp and paper mills or food-related industries such as meat processing, and industrial waste streams containing carbon molecules.

An exciting new technology, the membrane aerated biofilm reactor (MABR), refines this process to use 90% less energy for aeration. Air is gently blown into a spirally wound membrane in a tank, with air on one side of the membrane and mixed liquor on the other. Nitrification-denitrification is achieved by a biofilm that forms on the membrane. The result is an effluent suitable for irrigation or release into the environment.

Anaerobic Treatment

By contrast, anaerobic treatment uses bacteria to help organic material deteriorate in an oxygen-free environment. Lagoons and septic tanks may use anaerobic processes. The best-known anaerobic treatment is anaerobic digestion, which is used for treating food and beverage manufacturing effluents, as well as municipal wastewater, chemical effluent, and agricultural waste.

Anaerobic digestion produces biogas, which lets users create a source of income from waste.

Further Treatment

The type of biological treatment selected for wastewater treatment, whether aerobic or anaerobic, depends on a wide range of factors, including compliance with environmental regulations on discharge quality.

Biological treatments are often supplemented with treatments including chlorination and carbon filtration, as well as technologies like reverse osmosis and ultrafiltration.

Researchers continue to look for ways to optimize conventional biological wastewater treatment. In one example, Finnish researchers added iron sulfate to wastewater before biological treatment to reduce phosphorous in tough-to-treat pulp mill wastewater. Other researchers have used ultraviolet light to remove challenging substances such as chemical residues and pharmaceutical compounds. And, MABR's ground-breaking aeration model saves so much energy that it makes treatment possible in remote areas.

So, while biological treatment has a long history, it's continuing to evolve in ways that make it more effective, efficient, and available.

Biological Wastewater Treatment with Activated-sludge Process

The influent wastewater (e.g. municipal wastewater) goes through several stages in which different compound are removed out of the wastewater.

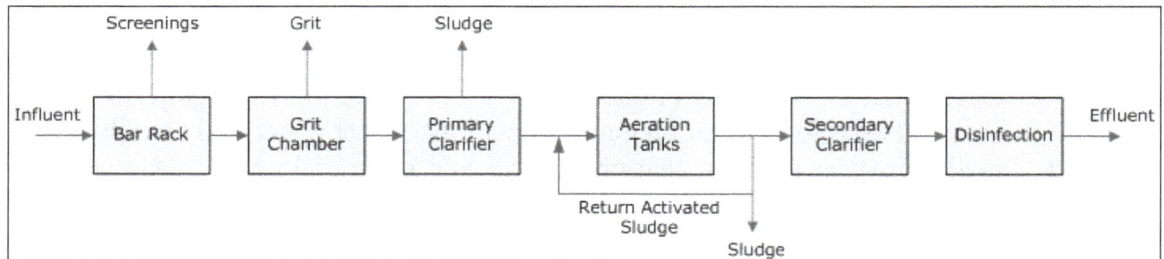

Simplified flow diagram for a biological wastewater treatment with a activated-sludge process.

In the Bar Rack coarse solids are removed, such as sticks, rags, and other debris in untreated wastewater by interception. By use of fine screening even floatable matter and algae are removed.

In the Grit Chamber grit is removed consisting of sand, gravel, cinders, or other heavy solid materials that have subsiding velocities or specific gravities substantially greater than those of the organic putrescible solids in wastewater. The Primary Clarifier is a basin where water has a certain retention time where the heavy organic solids can sediment (suspended solids). Efficiently designed and operated primary sedimentation tanks should remove from 50 to 70 percent of the suspended solids and 25 to 40 percent of the BOD. The influent of the aeration tank is mixed with activated sludge and in the Aeration Tank the mixed liquor is aerated. By aerating the mixed liquor the aerobic processes will be stimulated, the growth rate of bacteria will be must faster. Because the bacteria deplete the substrate, flocculation takes place . The soluble substrate becomes a solid biomass. These flocks of biomass will sediment in the Secondary Clarifier. At the end of the process the effluent water is treated to Disinfect it and make it free of disease-causing organisms.

Anaerobic Biological Wastewater Treatment

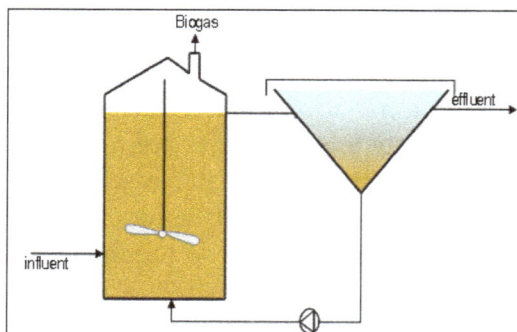

Anaerobic contact reactor.

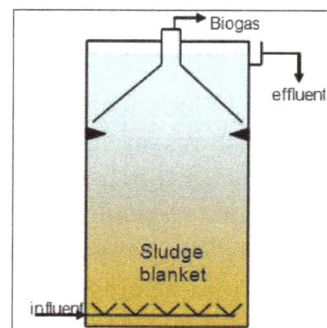

Upflow reactor.

Anaerobic treatments on wastewater are normally implemented when treating more concentrated wastewater. The anaerobic sludge contains various groups of micro organisms that work together to eventually convert organic material to biogas via hydrolysis and acidification. Biogas typically consists of 70% methane (CH_4) and 30% carbon dioxide (CO_2) with residual fractions of other gases (e.g. H_2 and H_2S). The methane can be used as an energy source. Anaerobic reactors can be implemented in a variety of ways.

The contact reactor is comparable with a conventional active sludge system, but under anaerobic conditions. The sludge is mixed with wastewater in the reactor and is then separated in the sedimentation tank and returned to the reactor.

In the anaerobic upflow reactor, the influent is introduced at the bottom of the vertical reactor. The sludge in the reactor is primarily grain shaped and forms a blanket in the reactor, with the most compact sludge grains at the bottom and the lighter grains and heavier sludge floccules above it. Very light sludge floccules will be released by the upward flow, but can potentially be collected in a sedimentation tank. The biogas is collected and disposed of at the top of the reactor, separately from the partly purified water and the sludge.

In addition to the contact reactor and the upflow reactor, other types are also available:

- Conventional digester:

This type is primarily implemented for the fermentation of RWZI sludge and liquid organic waste. The system is characterised by very low loads and a large volume in order to achieve the longest possible retention time. This type of reactor does not include recirculation of anaerobic sludge.

- Packed anaerobic filter (sludge on carrier):

This reactor is filled with carrier material and is normally used as an upflow reactor.

- UASB (upflow anaerobic sludge blanket) or EGSB (expanded granular sludge bed):

Both systems are variations of the upflow reactor. The main difference between the two is the increased recirculation of the EGSB reactor. Together with the prominent sludge grain, this enables higher loads in the EGSB (15-30 kg $COD/m^3/day$).

- Anaerobic membrane reactor:

This type of application uses membranes for sludge-water separation. To date, little use has been made of this system.

An extra purification phase will often be implemented after anaerobic purification, e.g. for the removal of residual fractions of COD and nutrients N and P. This often involves the use of an aerobic post-purification treatment.

Specific Advantages and Disadvantages

Advantages:

- Formation of biogas:

The encountered organic pollutant is converted into biogas with a high energetic value. This, for example, allows the energy needed to operate the water purification system to be fully or partly recuperated.

- High loads:

The volumetric load (COD load per m³ active volume per day) in an anaerobic reactor is typically 5 to 10 times higher than aerobic wastewater purification.

- Very little sludge production:

Sludge growth in an anaerobic reactor is 4-5 times lower than in an aerobic system.

Working in campaigns is possible. If the anaerobic sludge is not fed, it will hibernate , which means longer periods without food can be spanned without excessive sludge mortality. The system will almost immediately become active after re-start.

Disadvantages:

- Incomplete break-down of organic compounds: Need for post-purification via, for example, aerobic purification;

- No thorough nutrient removal: Again, later aerobic purification with nutrient removal is often needed;

- Most efficient purification in the mesophilic range, i.e. between 30-37°C, whereby the influent must be heated in most cases;

- Less robust system with regards to toxicity and inhibition;

- Risk of odour problems.

Application

Anaerobic purification is implemented in various sectors. In the food sector, this technique is regularly used to reduce the high cost of aerobic waste purification by partially breaking down the organic load and converting into biogas.

Anaerobic processes are also frequently used to ferment aerobic sludge and fluid organic waste.

Boundary Conditions

Anaerobic purification is often preceded by a tank for buffering and conditioning the wastewater. The typical retention time is two days. Acidification takes place up to pH 5.5 or 6, along with hydrolysis of the suspended matter.

Anaerobic reactors are typically used for wastewater purification when the wastewater complies with the following conditions:

- Average to high COD concentration.

- Temperatures from approximately 20°C.

- Average to low salt concentrations.

- Low sulphate concentration (ratio of COD concentration over S04 concentration less than 10).

- Low concentration of fats/oils.

- Absence of toxic components.

The effluent from anaerobic purification will normally not comply with applicable discharge norms and will require an extra purification step. Anaerobic post-purification will be selected in many cases, whereby the organic and nutrient load will be reduced.

Effectiveness

The anaerobic reactor can be implemented for removing the following parameters:

- COD: On average, the reactor will remove 80-90% of ingoing COD;

- N: Is incorporated into the sludge at a rate of 13g N per 1000g removed COD;

- P: Is incorporated into the sludge at a rate of 3g P per 1000g removed COD;

Approximately 0.35-0.4 Nm^3 biogas is produced per kg COD that is removed from the influent. The caloric value amounts to 20 to 30 MJ/Nm^3.

Support Aids

No supports aids are needed.

Environmental Issues

It is important to correctly collect and handle the formed biogas, in order to prevent it from escaping into the atmosphere. 60-75% of the biogas consists of methane, which is a greenhouse gas with an impact that is approximately 20 times greater than carbon dioxide.

Complexity

Despite the biological processes in the anaerobic reactor, this is a fairly simple system and is, in terms of complexity, comparable with conventional aerobic water purification.

Level of Automation

Anaerobic wastewater purification techniques can be fully automated, as can aerobic wastewater purification techniques.

Anaerobic Digestion of Sewage Sludge

Anaerobic digestion is one of the most widely used processes for the stabilisation of wastewater treatment plant sludge. Due to its capacity to reduce the amount of organic matter up to 50%, anaerobic digestion represents a necessary step of sludge treatment prior to drying and incineration, optimising the post-treatment process and saving costs. Furthermore, the generated biogas with a high proportion of methane can be used as an energy source.

Advantages

- Generation of CH4 as an additional source of energy.

- Reduction of organic matter, optimising further treatments.

- The remaining sludge can be used as soil conditioner.

- Reduces production of landfill methane, a greenhouse gas.

Disadvantages

- Accumulation of heavy metals and contaminants in sludge.

- High capital costs.

- Skilled manpower for design, construction and O&M.

- High complexity to maintain optimal reaction conditions.

Design and Construction Principles

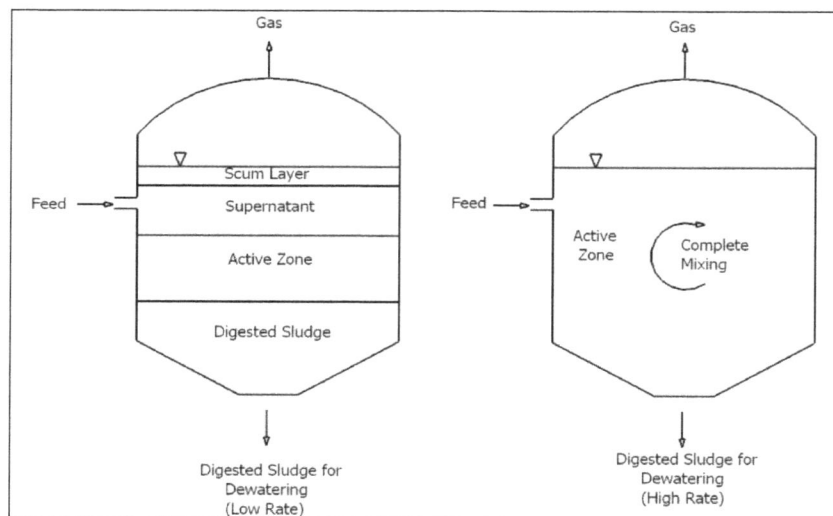

Design of an anaerobic sewage sludge digester.

Anaerobic digestion is the biological degradation of organic matter in the absence of free oxygen. During this process, much of the organic matter is converted to biogas (methane, carbon dioxide and water). Also known as methane fermentation or anaerobic sludge stabilisation, this process can reduce the organic matter content of sludge by 40 and 50 %. Two different types in anaerobic sludge digestion process are in practice: (1) Low rate digestion: a large storage tank, occasionally, with some heating facility, and (2) High rate digestion: with pre-thickening of raw sludge, complete mixing, heating and uniform feeding of raw sludge. Sludge feeding or organic loading rate (OLR) is expressed in terms of volatile solids (VS). Typically, it is 0.5– 0.6 kg VS/m³/day for low rate digestion and 3.2–7.2 kg VS/m³/day for high load digestion. The retention time is usually about one month. The HRT and the extent of each of the three reactions occurring during anaerobic digestion (hydrolysis, acidogenesis and methanogenesis) are directly related. The process can either be thermophilic digestion, in which sludge is fermented in tanks at a temperature of 55°C or mesophilic, at a temperature of around 36°C. It can be designed with a batch or continuous configuration, in either one or two stages. The biodigester is an air and watertight structure that provides anaerobic conditions.

Operation and Maintenance

The O&M of an anaerobic digester requires a strict organisation and the continuous involvement of experts. There are many problems associated with its operation that should be given special attention: over-pumping of raw sludge, excessive withdrawal of the digested sludge, foaming, maintenance of an optimum/uniform temperature and inadequate mixing. In order to maintain optimal conditions for the microbial population to growth, nutrients (N and P) should be added in the form of ammonium chloride, aqueous ammonia, urea phosphate salts and phosphoric acid.

Cost Considerations

The capital requirements to install a digester vary depending on the design chosen, size and choice of equipment for utilisation of the biogas. The cost per unit volume installed increases for smaller treatment plants. Costs based on the price level in 2006 ranged from 250 USD/m³ (for a plant of capacity 200,000 PE) up to 1,000 USD/m³ (for 25,000 PE) (VAN HAANDEL & VAN DER LUBBE 2012). The annual O&M requirements include labour hours (skilled and unskilled), electrical energy, fuel energy as well as the annual materials and maintenance. Thermophilic processes require heating systems, increasing the O&M costs.

Anaerobic digesters to treat excess sludge were first implemented in Exeter, UK in 1885, and since then the technology and its benefits have been well known for more than 100 years (PETITPAIN 2013). Because of the incertitude linked to the use of the remaining sludge in agriculture, this treatment solution remained weak in the past decades meanwhile other technologies, such as sludge drying, incineration and gasification were preferred. Today, due to the increasing energy costs and the introduction of green electricity certification in countries such as Germany and Belgium, this technology is gaining terrain in Europe. In North America, anaerobic digestion is the dominant municipal sludge stabilisation technology. The anaerobic digestion process has been generally used for WWTPs having wastewater flow less than 4,000 m³/day to more than 757,000 m³/day. Despite the fact that the production of electricity from the digested gas recovery becomes more cost effective for plants with daily flows greater than 38,000 m³/d, it is the preferable choice for WWTPs capacities less than 10,000 inhabitants in Germany.

Role of Microorganisms used in Wastewater Treatment

Wastewater can be detrimental to the environment if left untreated. That's because waste from humans and pets are a source of several types of waterborne diseases and bacterial contamination.

Thanks in part to microorganisms, treating wastewater and sewage is possible. The role of microorganisms in wastewater treatment helps to treat and purify wastewater and make it less harmful to the environment.

While there are many different microbes used in sewage treatment, there are three well-known microbes that play an instrumental role in keeping sewage clean. Each of these types of bacteria help the treatment process in a unique way to ensure there is little to no impact on the surrounding environment.

Common Microorganisms used in Wastewater Treatment

Here is a list of bacteria used in sewage treatment you can reference:

Aerobic Bacteria

Aerobic bacteria are mostly used in new treatment plants in what is known as an aerated environment. This bacterium uses the free oxygen within the water to degrade the pollutants in the wastewater and then converts it into energy that it can use to grow and reproduce.

For this type of bacteria to be used correctly, it must have oxygen added mechanically. This will ensure the bacteria are able to do their job correctly and continue to grow and reproduce on its food source.

Anaerobic Bacteria

Anaerobic bacteria are used in wastewater treatment on a normal basis. The main role of these bacteria in sewage treatment is to reduce the volume of sludge and produce methane gas from it.

The great thing about this type of bacteria and why it's used more frequently than aerobic bacteria is that the methane gas, if cleaned and handled properly, can be used as an alternative energy source. This is a huge benefit considering the already high wastewater treatment energy consumption levels.

Unlike aerobic bacteria, this type of bacteria is able to get more than enough oxygen from its food source and will not require adding oxygen to help do its job. Phosphorus removal from wastewater is another benefit of anaerobic microbes used in sewage treatment.

Facultative

Facultative microorganisms in sewage treatment are bacteria that can change between aerobic and anaerobic depending on the environment they are in. Note that these bacteria normally prefer to be in an aerobic condition.

Many industrial and municipal wastewater treatment plants use bacteria and other microorganisms to help with the process of cleaning sewage. Picking the right bacteria can be tricky since your selection depends on the condition of your area for effective use. Wastewater treatment can also provide a great source for alternative energy if the anaerobic bacteria are handled correctly.

Learning the names of microbes used in sewage treatment and the role bacteria in sewage treatment plays doesn't have to be a solo job. Consider the water treatment solutions available from AOS to learn more about the role of microorganisms in water treatment and how microorganisms in the wastewater treatment process can help keep your water healthy.

Microorganisms in Activated Sludge

In the activated sludge process, microorganisms are mixed with wastewater. The microorganisms come in contact with the biodegradable materials in the wastewater and consume them as food. In addition, the bacteria develop a sticky layer of slime around the cell wall that enables them to clump together to form bio-solids or sludge that is then separated from the liquid phase. The successful removal of wastes from the water depends on how efficiently the bacteria consume the organic material and on the ability of the bacteria to stick together, form floc, and settle out of the bulk fluid. The flocculation (clumping) characteristics of the microorganisms inactivated sludge enable them to amass to form solid masses large enough to settle to the bottom of the settling basin. As the flocculation characteristics of the sludge improves, so is the improved settling and improved wastewater treatment.

After the aeration basin, the mixture of microorganisms and wastewater (mixed liquor) flows into a settling basin or clarifier where the sludge is allowed to settle. Some of the sludge volume is continuously recirculated from the clarifier, as Returned Activated Sludge (RAS), back to the aeration basin to ensure adequate amounts of microorganisms are maintained in the aeration tank. The microorganisms are again mixed with incoming wastewater where they are reactivated to consume organic nutrients. Then the process starts again.

The activated sludge process, under proper conditions, is very efficient. It removes 85 to 95 percent of the solids and reduces the biochemical oxygen demand (BOD) about the same amount. The efficiency of this system depends on many factors, including wastewater climate and characteristics. Toxic wastes that enter the treatment system can disrupt the biological activity. Wastes heavy in soaps or detergents can cause excessive frothing and thereby create aesthetic or nuisance problems. In areas where industrial and sanitary wastes are combined, industrial wastewater must often be pretreated to remove the toxic chemical components before it is discharged into the activated sludge treatment process. Nevertheless, microbiological treatment of wastewater is by far the most natural and effective process for removing wastes from water.

There are five major groups of microorganisms generally found in the aeration basin of the activated sludge process:

- Bacteria: Aerobic bacteria remove organic nutrients.

- Protozoa: Remove & digests dispersed bacteria and suspended particles.

- Metazoa: Dominate longer age systems including lagoons.

- Filamentous bacteria: Bulking sludge (poor settling & turbid effluent).

- Algae and fungi: Fungi is present with pH changes & older sludge.

Bacteria are primarily responsible for removing organic nutrients from the wastewater.

Protozoa play a critical role in the treatment process by removing and digesting free swimming dispersed bacteria and other suspended particles. This improves the clarity of the wastewater effluent. Like bacteria, some protozoa need oxygen, some require very little oxygen, and a few can survive without oxygen.

The types of protozoa present give us some indication of treatment system performance which are classified as follows:

- Amoebae: Little effect on treatment & die off as amount of food decreases.

- Flagellates: Feed primarily on soluble organic nutrients.

- Ciliates: Clarify water by removing suspended bacteria.

    ◦ Ciliates: Free-swimming-Removes free-dispersed bacteria.

    ◦ Ciliates: Crawling (grazing)-Dominate activated sludge/good treatment.

    ◦ Ciliates: Stalked (sessile)-Dominates at process end.

Metazoa are multi-cellular organisms which are larger than most protozoa and have very little to do with the removal of organic material from the wastewater. Although they do eat bacteria, they also feed on algae and protozoa. A dominance of metazoa is usually found in longer age systems; namely, lagoon treatment systems. Although their contribution in the activated sludge treatment system is small, their presence does indicate treatment system conditions.

Three most common metazoa found in the activated sludge treatment system:

- Rotifers: Clarify effluent & are first affected by toxic loads.

- Nematodes: Feed on bacteria, fungi, small protozoa & other nematodes.

- Tardigrades (water bear): Survive environmental extremes & toxic sensitivity.

Filamentous bacteria are present when operational conditions drastically change. These bacteria grow in long filaments begin to gain an advantage. Changes in temperature, pH, DO, sludge age, or even the amounts of available nutrients such as nitrogen, phosphorus, oils & grease can affect these bacteria. The dominance of filamentous bacteria in the activated sludge treatment system can cause problems with sludge settling. At times excessive numbers of filamentous microorganisms interfere with floc settling and the sludge becomes bulky. This bulking sludge settles poorly and leaves behind a turbid effluent. Some filamentous microorganisms may cause foaming in the aeration basin and clarifiers.

Algae and fungi which are photosynthetic organisms and generally do not cause problems in activated sludge treatment systems, however there presence in the treatment system usually indicate problems associated pH changes and older sludge.

Problems with Biological Wastewater Treatment Systems

Biological wastewater treatment systems can be a big investment for your facility. When implemented correctly, these systems will provide your process with a long list of benefits, but there can be some challenges with purchasing these systems, too.

If your facility is considering biological wastewater treatment, you might be wondering if there are common problems with these systems and the best ways to avoid them.

In general, when engineering, designing, installing, and running a biological wastewater treatment systemfor your industrial facility, issues most commonly arise during the engineer/design and operational phases.

Common Engineering and Design Issues

Most problems with biological wastewater treatment systems occur at the engineering and design-stages. When studies aren't completed for long enough periods of time over vast-enough amounts of data, it's common for a system to fall short in some way.

Choosing Inappropriate Treatment Technologies

Sometimes chemical solutions might be a better selection for the application rather than biological. We see this a lot when harmful components of the wastewater intoxicate the bacteria or if there is a very low degradable but high nonbiodegradable organic content. In these cases, chemical oxidation is usually the preferred treatment option versus biological.

It's important for your water treatment specialists to look at the ratio of BOD to COD to ensure the waste is, in fact, biodegradable and there is not much toxicity. Dairy products, cheese, and whey, etc., produce highly biodegradable waste, in which biological treatment technologies are a good fit. But when BOD to COD ratios are low, this indicates there are a lot of organics present that *aren't* biodegradable and that physical/chemical technologies might be a better choice.

Gathering Inaccurate Waste Stream Data

True of any treatment system design, it's important to gather as much information about your waste stream as possible at the very beginning of the engineering process. Do you have the right flow rates? How are you sure they're accurate? Flow rates in food processing and dairy facilities, among others, go up and down from day to day and change over time. Often, we see these facilities spend two or three days keeping track of their wastewater data when they might need a thorough, two-week study instead. This ensures the system designers and engineers not only know the average rates but are certain of the maximums and minimums over time.

What about the chemistry and chemical makeup of the water? Is there a good level of data about your normal average concentration of BOD, COD, sugar, fats, oils, solids, pH, temperature, and salinity? Is there a strong understanding of the chemistry of the water?

This is the essential information that will be factored into your system's design, and with biological wastewater treatment, you don't just need a good design basis, but you'll also need to know the necessary pretreatment measures for the water. Pretreatment could be as simple as equalization or solids removal, but you'll often need pH and temperature control capabilities, too.

If these steps aren't designed correctly according to your facility's true parameters, you will likely see an increase of operational problems with your plant after implementing the new system.

Improper Aeration Capabilities

When designing an effective biological water treatment system, it's important to not over or under aerate the water. For example, if the system is anaerobic or anoxic (where no oxygen is used), the design must enable your facility to create that anoxic, oxygen-free environment or the anaerobic reducing environment for the proper bacteria to grow.

When the system is under aerated, itruns out of oxygen, and the first thing you're going to notice is foul odors. If you over aerate the system, it won't create any operational issues, per se, but you'll be running rotary aeration equipment unnecessarily, which will consume more electricity than you need and inflate costs. For a facility that processes a million gallons of waste per day, it would be easy to accrue hundreds of thousands of dollars in excess operating cost per year because the aeration is overdesigned.

Getting the Correct pH

If the pH component of your system isn't designed correctly, bacteria stop growing, which means they won't remove the necessary pollutants quickly enough and you'll have those pollutants breaking through the system and going out in the effluent. This means you won't meet your effluent limits and could encounter surcharges or fines. You may even have your plant shut down if you're discharging to a creek or a river, and you have excess pollutants in your effluent because you're not controlling the pH.

How to Avoid these Common Engineering and Design Issues

These issues can have a big impact on your biological wastewater treatment system, so how do you avoid them?

- Wastewater characterization study: By pulling samples and sending them to the laboratory, you're making sure the proper parameters are analyzed that are necessary for developing the design. To do this, repeatedly measure flow throughout the day along all the major shifts of the plant—and do that for several days to cover all the variations. Then pull a report together, and from that data, your water treatment experts will form the basis of the system's technology and configuration choices.

- Laboratory study: A proper laboratory or wastewater treatability study can take anywhere from three to five weeks, but this step is extremely important. It will illustrate whether there is toxicity; how much of the BOD will need to be taken out; if bacteria can be used to treat the water all the way to the discharge limits or, if it can't be treated all the way, what steps are needed afterward (posttreatment technologies) to get to the discharge limits. These studies are extremely helpful for avoiding design problems when you grow biological organisms in the wastewater. It will help your facility design the pH, understand the source of any odors, know if there's a need for extra nutrients and chemical dosing of nitrogen or phosphorous, and everything else about bacterial growth on the wastewater pollutants.

- Pilot study: If, after the wastewater characterization and laboratory studies are complete and it's decided you're working with an extremely complex waste with lots of variation, you may need to run a pilot plant study at the facility for anywhere from two to five months. This will put the biological technology, pretreatment, posttreatment, aeration, pH control, temperature control, nutrient dosing, etc., into practice. By running a slipstream into a small pilot at the flows and loading rates that duplicate the full-scale commercial design, all the variation from day to day can be properly analyzed. This step not only demonstrates how the technology will work, it also generates engineering data, which will help your water treatment engineers design the optimal system for the biological water treatment. With a successful pilot plant study, your system-design engineers should be able to guarantee the full-scale commercial system that will be the best fit for your waste stream.

Common Plant Operations Issues

Although most problems with biological wastewater treatment systems typically occur in the design and engineering phases, some also occur during the operation of the system.

Let's say the design, construction, and installation of the system is complete. It's built and ready to be started up. In order to facilitate a successful implementation at the plant, there needs to be a dedicated team of operators available for training. The engineers who designed the system should provide an operation manual, and some will even include very specific, detailed SOPs which are written so the operators how to make up the nutrient solution. They should have some type of skills to operate the engineered system as they will need to learn about growing and maintaining living organisms, and if the necessary life support chemistry isn't followed, the bacteria will die.

The systems can't be overloaded and exceed the capacity with either the volume of water or the load of BOD/COD of the pollutant that the system is designed to remove without having problems with the effluent. So it's important to have operational procedures with contingencies in place to always keep your wastewater treatment plant up and running at specification within limits in order to discharge.

They'll also need to maintain the biological system by collecting daily, weekly, and monthly samples and doing the analytical work and maintaining them in a spreadsheet database so loss of performance can be detected.

It's also important to maintain a relationship with your water treatment specialists and system manufacturer because they designed your system and will be the best consultant for solving problems before they occur.

References

- What-is-biological-wastewater-treatment: fluencecorp.com, Retrieved 18 June, 2019

- Anaerobic-biological-wastewater-treatment: emis.vito.be, Retrieved 17 January, 2019

- Anaerobic-digestion-of-sewage-sludge, module-1-nawatech-basics: sswm.info, Retrieved 09 June, 2019

- Role-microbes-microorganisms-used-wastewater-sewage-treatment: aosts.com, Retrieved 25 August, 2019

- Microorganisms-in-activated-sludge: watertechonline.com, Retrieved 23 July, 2019

- Common-problems-with-biological-wastewater-treatment-systems-how-to-avoid-them: samcotech.com, Retrieved 19 May, 2019

Chapter 5

Industrial Wastewater Treatment

Industrial Wastewater Treatment is defined as the process which are used in treatment of wastewater produced by industries. It includes wastewater treatment in fertilizer industry, iron and steel industry, cement and ceramic industry, paper and pulp industry, etc. All the concepts related to industrial wastewater treatment have been carefully analyzed in this chapter.

Industrial wastewater treatment covers the mechanisms and processes used to treat waters that have been contaminated in some way by anthropogenic industrial or commercial activities prior to its release into the environment or its re-use.

Most industries produce some wet waste although recent trends in the developed world have been to minimise such production or recycle such waste within the production process. However, many industries remain dependent on processes that produce wastewaters.

Sources of Industrial Wastewater

Agricultural Waste

Breweries:

Beer is a fermented beverage with low alcohol content made from various types of grain. Barley predominates, but wheat, maize, and other grains can be used. The production steps include:

- Malt production and handling: Grain delivery and cleaning; steeping of the grain in water to start germination; growth of rootlets and development of enzymes (which convert starch into maltose); kilning and polishing of the malt to remove rootlets; storage of the cleaned malt.

- Wort production: Grinding the malt to grist; mixing grist with water to produce a mash in the mash tun; heating of the mash to activate enzymes; separation of grist residues in the lauter tun to leave a liquid wort; boiling of the wort with hops; separation of the wort from the trub/hot break (precipitated residues), with the liquid part of the trub being returned to the lauter tub and the spent hops going to a collection vessel; and cooling of the wort.

- Beer production: Addition of yeast to cooled wort; fermentation; separation of spent yeast by filtration, centrifugation or settling; bottling or kegging.

- Water consumption for breweries generally ranges 4–8 cubic meter per cubic meter (m3/m3) of beer produced.

Breweries can achieve an effluent discharge of 3–5 m3/m3 of sold beer (exclusive of cooling waters). Untreated effluents typically contain sus-pended solids in the range 10–60 milligrams per liter (mg/l), biochemical oxygen demand (BOD) in the range 1,000–1,500 mg/l, chemical oxygen demand (COD) in the range 1,800–3,000 mg/l, and nitrogen in the range 30–100 mg/l. Phosphorus can also be present at concentrations of the order of 10–30 mg/l. Effluents from individual process steps are variable. For example, bottle washing produces a large volume of effluent that, however, contains only a minor part of the total organics discharged from the brewery. Effluents from fermentation and filtering are high in organics and BOD but low in volume, accounting for about 3% of total wastewater volume but 97% of BOD. Effluent pH averages about 7 for the combined effluent but can fluctuate from 3 to 12 depending on the use of acid and alkaline cleaning agents. Effluent temperatures average about 30°C.

Dairy Industry

The dairy industry involves processing raw milk into products such as consumer milk, butter, cheese, yogurt, condensed milk, dried milk (milk powder), and ice cream, using processes such as chilling, pasteurization, and homogenization. Typical by-products include buttermilk, whey, and their derivatives.

Waste Characteristics Dairy effluents contain dissolved sugars and proteins, fats, and possibly residues of additives. The key parameters are biochemical oxygen demand (BOD), with an average ranging from 0.8 to 2.5 kilograms per metric ton (kg/t) of milk in the untreated effluent; chemical oxygen demand (COD), which is normally about 1.5 times the BOD level; total suspended solids, at 100–1,000 milligrams per liter (mg/l); total dissolved solids: phosphorus (10–100 mg/l), and nitrogen (about 6% of the BOD level). Cream, butter, cheese, and whey production are major sources of BOD in wastewater. The waste load equivalents of specific milk constituents are: 1 kg of milk fat = 3 kg COD; 1 kg of lactose = 1.13 kg COD; and 1 kg protein = 1.36 kg COD. The wastewater may contain pathogens from contaminated materials or production processes. A dairy often generates odors and, in some cases, dust, which need to be controlled. Most of the solid wastes can be processed into other products and byproducts.

Pulp and Paper Industry

The pulp and paper industry is one of worlds oldest and core industrial sector. The socio-economic importance of paper has its own value to the country's development as it is directly related to the industrial and economic growth of the country. Paper manufacturing is a highly capital, energy and water intensive industry. It is also a highly polluting process and requires substantial investments in pollution control equipment. The pulp and paper mill is a major industrial sector utilizing a huge amount of lignocellulosic materials and water during the manufacturing process, and releases chlorinated lignosulphonic acids, chlorinated resin acids, chlorinated phenols and chlorinated hydrocarbons in the effluent. About 500 different chlorinated organic compounds have been identified including chloroform, chlorate, resin acids, chlorinated hydrocarbons, phenols, catechols, guaiacols, furans, dioxins, syringols, vanillins, etc. These compounds are formed as a result of reaction between residual lignin from wood fibres and chlorine/chlorine compounds used for bleaching. Colored compounds and Adsorbable Organic Halogens (AOX) released from pulp and paper mills into the environment poses numerous problems. The wood pulping and production of the paper products generate a considerable amount of pollutants characterized by Biochemical Oxygen Demand (BOD), Chemical Oxygen Demand (COD), Suspended Solids (SS), toxicity, and colour when untreated or poorly treated effluents are discharged to receiving waters. The effluent is toxic to aquatic organisms and exhibits strong mutagenic effects and physiological impairment.

Iron and Steel Industry

The production of iron from its ores involves powerful reduction reactions in blast furnaces. Cooling waters are inevitably contaminated with products especially ammonia and cyanide. Production of coke from coal in coking plants also requires water cooling and the use of water in by-products separation. Contamination of waste streams includes gasification products such as benzene, naphthalene, anthracene, cyanide, ammonia,phenols, cresols together with a range of more complex organic compounds known collectively as polycyclic aromatic hydrocarbons (PAH).

The conversion of iron or steel into sheet, wire or rods requires hot and cold mechanical transformation stages frequently employing water as a lubricant and coolant. Contaminants include hydraulic oils, tallow and particulate solids. Final treatment of iron and steel products before onward sale into manufacturing includes pickling in strong mineral acid to remove rust and prepare the surface for tin or chromium plating or for other surface treatments such as galvanisation or painting. The two acids commonly used are hydrochloric acid and sulfuric acid. Wastewaters include acidic rinse waters together with waste acid. Although many plants operate acid recovery plants, (particularly those using Hydrochloric acid), where the mineral acid is boiled away from the iron salts, there remains a large volume of highly acid ferrous sulfate or ferrous chloride to be disposed of. Many steel industry wastewaters are contaminated by hydraulic oil also known as soluble oil.

Mines and Quarries

The principal Wastewaters associated with mines and quarries are slurries of rock particles in water. These arise from rainfall washing exposed surfaces and haul roads and also from rock washing and grading processes. Volumes of water can be very high, especially rainfall related arisings on large sites. Some specialized separation operations, such as coal washing to separate coal from native rock using density gradients, can produce wastewater contaminated by fine particulate haematite and surfactants. Oils and hydraulic oils are also common contaminants. Wastewater from metal mines and ore recovery plants are inevitably contaminated by the minerals present in the native rock formations. Following crushing and extraction of the desirable materials, undesirable materials may become contaminated in the wastewater. For metal mines, this can include unwanted metals such as zinc and other materials such as arsenic. Extraction of high value metals such as gold and silver may generate slimes containing very fine particles in where physical removal of contaminants become particularly difficult.

Food Industry

Wastewater generated from agricultural and food operations has distinctive characteristics that set it apart from common municipal wastewater managed by public or private wastewater treatment plants throughout the world: it is biodegradable and nontoxic, but that has high concentrations of biochemical oxygen demand (BOD) and suspended solids (SS). The constituents of food and agriculture wastewater are often complex to predict due to the differences in BOD and pH in effluents from vegetable, fruit, and meat products and due to the seasonal nature of food processing and postharvesting.

Processing of food from raw materials requires large volumes of high grade water. Vegetable washing generates waters with high loads of particulate matter and some dissolved organics. It may also contain surfactants.

Animal slaughter and processing produces very strong organic waste from body fluids, such as blood, and gut contents. This wastewater is frequently contaminated by significant levels of antibiotics and growth hormonesfrom the animals and by a variety of pesticides used to control external parasites. Insecticide residues in fleeces is a particular problem in treating waters generated in wool processing.

Processing food for sale produces wastes generated from cooking which are often rich in plant organic material and may also contain salt, flavourings, colouring material and acids or alkali. Very significant quantities of oil or fats may also be present.

Complex Organic Chemicals Industry

A range of industries manufacture or use complex organic chemicals. These include pesticides, pharmaceuticals, paints and dyes, petro-chemicals, detergents, plastics, paper pollution, etc. Waste waters can be contaminated by feed-stock materials, by-products, product material in soluble or particulate form, washing and cleaning agents, solvents and added value products such as plasticisers.

Nuclear Industry

The waste production from the nuclear and radio-chemicals industry is dealt with as Radioactive waste.

Water Treatment

Water treatment for the production of drinking water is dealt with elsewhere. Many industries have a need to treat water to obtain very high quality water for demanding purposes. Water treatment produces organic and mineral sludges from filtration and sedimentation. Ion exchange using natural or synthetic resins removes calcium, magnesium and carbonate ions from water, replacing them with hydrogen and hydroxyl ions. Regeneration of ion exchange columns with strong acids and alkalis produces a wastewater rich in hardness ions which are readily precipitated out, especially when in admixture with other wastewaters.

Treatment of Industrial Wastewater

The different types of contamination of wastewater require a variety of strategies to remove the contamination.

Solids Removal

Most solids can be removed using simple sedimentation techniques with the solids recovered as slurry or sludge. Very fine solids and solids with densities close to the density of water pose special problems. In such case filtration or ultrafiltration may be required. Although, flocculation may be used, using alum salts or the addition of polyelectrolytes.

Oils and Grease Removal

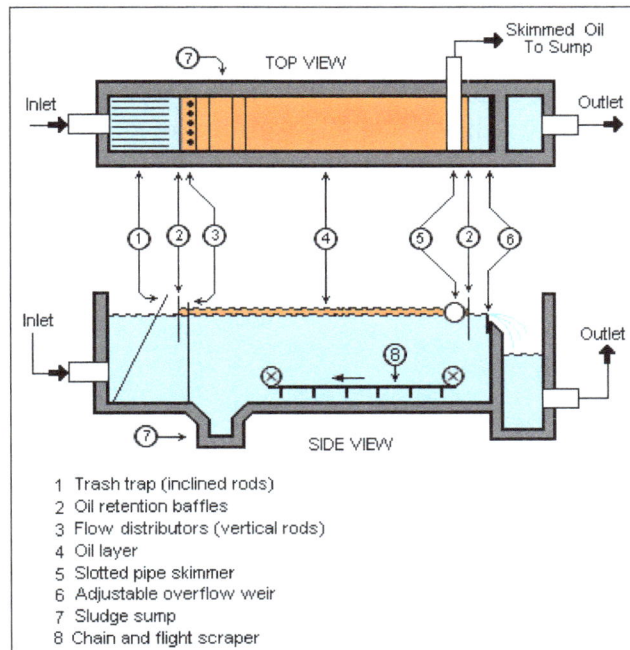

A typical API oil-water separator used in many industries.

Many oils can be recovered from open water surfaces by skimming devices. Considered a dependable and cheap way to remove oil, grease and other hydrocarbons from water, oil skimmers can sometimes achieve the desired level of water purity. At other times, skimming is also a cost-efficient method to remove most of the oil before using membrane filters and chemical processes. Skimmers will prevent filters from blinding prematurely and keep chemical costs down because there is less oil to process.

Because grease skimming involves higher viscosity hydrocarbons, skimmers must be equipped with heaters powerful enough to keep grease fluid for discharge. If floating grease forms into solid clumps or mats, a spray bar, aerator or mechanical apparatus can be used to facilitate removal.

However, hydraulic oils and the majority of oils that have degraded to any extent will also have a soluble or emulsified component that will require further treatment to eliminate. Dissolving or emulsifying oil using surfactants or solvents usually exacerbates the problem rather than solving it, producing wastewater that is more difficult to treat.

The wastewaters from large-scale industries such as oil refineries, petrochemical plants, chemical plants, and natural gas processing plants commonly contain gross amounts of oil and suspended solids. Those industries use a device known as an API oil-water separator which is designed to separate the oil and suspended solids from their wastewater effluents. The name is derived from the fact that such separators are designed according to standards published by the American Petroleum Institute (API).

The API separator is a gravity separation device designed by using Stokes Law to define the rise velocity of oil droplets based on their density and size. The design is based on the specific gravity difference between the oil and the wastewater because that difference is much smaller than the specific gravity difference between the suspended solids and water. The suspended solids settles to the bottom of the separator as a sediment layer, the oil rises to top of the separator and the cleansed wastewater is the middle layer between the oil layer and the solids.

Typically, the oil layer is skimmed off and subsequently re-processed or disposed of, and the bottom sediment layer is removed by a chain and flight scraper (or similar device) and a sludge pump. The water layer is sent to further treatment consisting usually of an Electroflotation module for additional removal of any residual oil and then to some type of biological treatment unit for removal of undesirable dissolved chemical compounds.

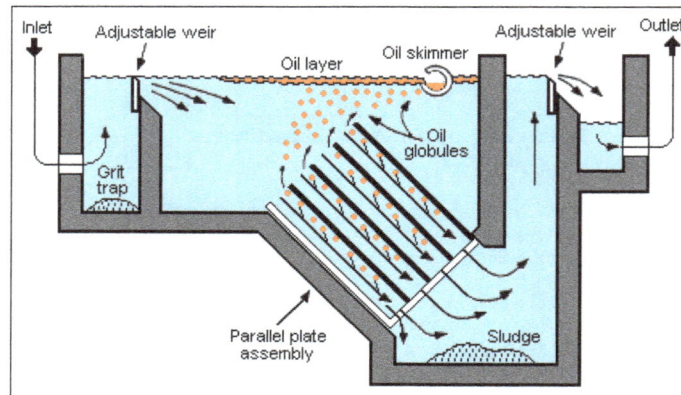

A typical parallel plate separator.

Parallel plate separators are similar to API separators but they include tilted parallel plate assemblies (also known as parallel packs). The parallel plates provide more surface for suspended oil droplets to coalesce into larger globules. Such separators still depend upon the specific gravity between the suspended oil and the water. However, the parallel plates enhance the degree of oil-water separation. The result is that a parallel plate separator requires significantly less space than a conventional API separator to achieve the same degree of separation.

Removal of Biodegradable Organics

Biodegradable organic material of plant or animal origin is usually possible to treat using extended conventional wastewater treatment processes such as activated sludge or trickling filter. Problems can arise if the wastewater is excessively diluted with washing water or is highly concentrated such as neat blood or milk. The presence of cleaning agents, disinfectants, pesticides, or antibiotics can have detrimental impacts on treatment processes.

Activated Sludge Process

Activated sludge is a biochemical process for treating sewage and industrial wastewater that uses air (or oxygen) and microorganisms to biologically oxidize organic pollutants, producing a waste sludge (or floc) containing the oxidized material. In general, an activated sludge process includes:

An aeration tank where air (or oxygen) is injected and thoroughly mixed into the wastewater. A settling tank (usually referred to as a "clarifier" or "settler") to allow the waste sludge to settle. Part of the waste sludge is recycled to the aeration tank and the remaining waste sludge is removed for further treatment and ultimate disposal.

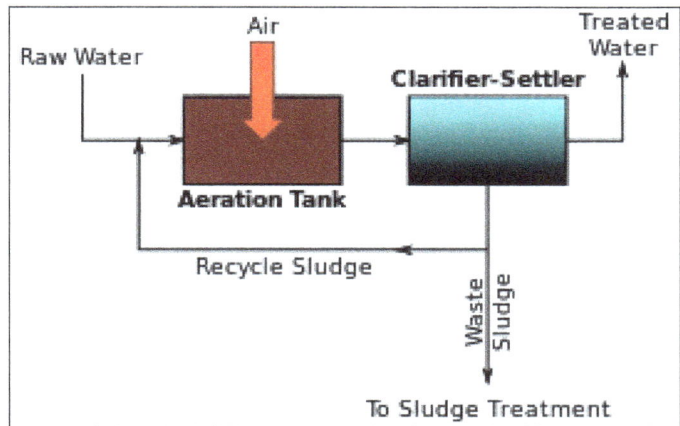

A generalized, schematic diagram of an activated sludge process.

Trickling Filter Process

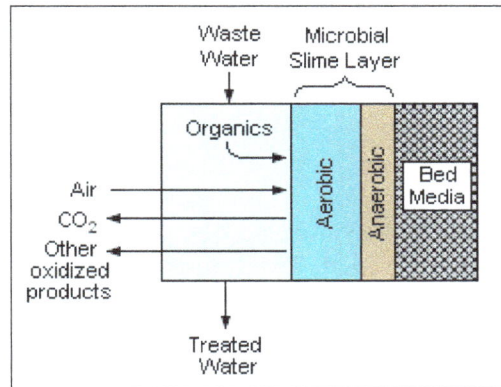

A schematic cross-section of the contact face of the bed media in a trickling filter.

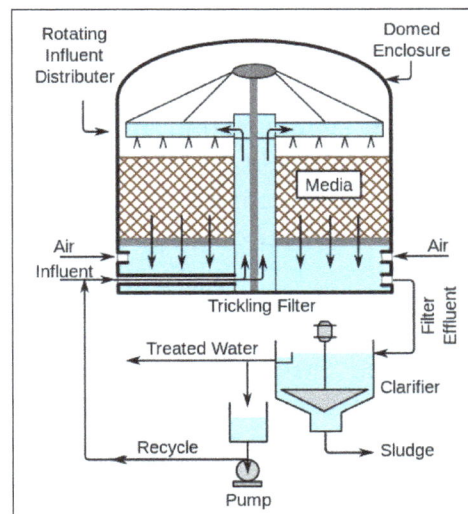

A trickling filter consists of a bed of rocks, gravel, slag, peat moss, or plastic media over which wastewater flows downward and contacts a layer (or film) of microbial slime covering the bed media. Aerobic conditions are maintained by forced air flowing through the bed or by natural convection of air. The process involves adsorption of organic compounds in the wastewater by the microbial slime layer, diffusion of air into the slime layer to provide the oxygen required for the biochemical oxidation of the organic compounds. The end products include carbon dioxide gas, water and other products of the oxidation. As the slime layer thickens, it becomes difficult for the air to penetrate the layer and an inner anaerobic layer is formed.

The components of a complete trickling filter system are: fundamental components:

- A bed of filter medium upon which a layer of microbial slime is promoted and developed.

- An enclosure or a container which houses the bed of filter medium.

- A system for distributing the flow of wastewater over the filter medium.

- A system for removing and disposing of any sludge from the treated effluent.

The treatment of sewage or other wastewater with trickling filters is among the oldest and most well characterized treatment technologies.

A trickling filter is also often called a trickle filter, trickling biofilter, biofilter, biological filter or biological trickling filter.

Treatment of other Organics

Synthetic organic materials including solvents, paints, pharmaceuticals, pesticides, coking products and so forth can be very difficult to treat. Treatment methods are often specific to the material being treated. Methods include Advanced Oxidation.

Processing, distillation, adsorption, vitrification, incineration, chemical immobilisation or landfill disposal. Some materials such as some detergents may be capable of biological degradation and in such cases, a modified form of wastewater treatment can be used.

Treatment of Acids and Alkalis

Acids and alkalis can usually be neutralised under controlled conditions. Neutralisation frequently produces a precipitate that will require treatment as a solid residue that may also be toxic. In some cases, gasses may be evolved requiring treatment for the gas stream. Some other forms of treatment are usually required following neutralisation.

Waste streams rich in hardness ions as from de-ionisation processes can readily lose the hardness ions in a buildup of precipitated calcium and magnesium salts. This precipitation process can cause severe furring of pipes and can, in extreme cases, cause the blockage of disposal pipes. A 1 metre diameter industrial marine discharge pipe serving a major chemicals complex was blocked by such salts in the 1970s. Treatment is by concentration of de-ionisation waste waters and disposal to landfill or by careful pH management of the released wastewater.

Treatment of Toxic Materials

Toxic materials including many organic materials, metals (such as zinc, silver, cadmium, thallium, etc.) acids, alkalis, non-metallic elements (such as arsenic or selenium) are generally resistant to biological processes unless very dilute. Metals can often be precipitated out by changing the pH or by treatment with other chemicals. Many, however, are resistant to treatment or mitigation and may require concentration followed by landfilling or recycling. Dissolved organics can be incinerated within the wastewater by Advanced Oxidation Processes.

Types of Industrial Wastewater Equipment

Wastewater treatment is an increasingly critical topic of discussion that has been addressed at the highest levels of government and major corporations. To find a sustainable approach, companies can employ many different strategies to help themselves go beyond mere compliance and begin the process of improving global water quality.

More than 80 percent of all the wastewater from industry, homes, cities and agriculture flows back into the ecosystem via lakes, rivers and other bodies of surface water. This process repeats every day across the planet, polluting the environment while losing valuable nutrients and other recoverable materials in the process.

Each year in March, World Water Day serves as a reminder from the United Nations that a daily commitment is necessary for the successful reduction and reuse of wastewater. Guy Ryder, the director-general of the UN International Labor Organization (ILO) and the chairperson of UN-Water, believes that there must be a commitment to improve management of wastewater from the business community and the general public to make a difference.

From a strictly monetary standpoint, it makes sense for companies to adopt a formal wastewater treatment and reuse policy because it allows them to dramatically cut rising operational costs while increasing profitability. Disposing of spent water-based coolant or wash water is expensive. Companies must pay for handling, trucking and treatment by their local publicly operated treatment works (POTW). Adding to the expense is the clean water required to replace the initial volume.

Water disposal costs can vary based on:

- Local water supplies,

- Fuel prices,

- Trucking prices,

- Edicts of the POTW.

The obvious goal should always be to recycle coolant, wash water and other fluids internally. This will increase tool life, improve product quality, reduce maintenance, and prolong the usage of working fluids. Having a fluid-recycling process in place means when the time comes to dispose of wastewater and fluids, companies will have a lower volume to discard or a concentrated stream they can treat themselves for lower cost handling at the POTW.

While it certainly makes business sense to implement such a process, for companies that are unfamiliar with the treatment and reuse of in-house wastewater, it can appear to be a daunting task.

On the surface, up-front costs often associated with adding a wastewater treatment system can seem prohibitive.

At the heart of any system is the equipment. For companies concerned about the effect their wastewater has on the environment and their bottom lines, a variety of options are available – each designed to perform specific types of treatment and deliver a quick return on investment (ROI).

The following list includes the seven most common types of wastewater equipment, how each operates, and how it affects a business's profitability:

Ultrafiltration Systems

Ultrafiltration (UF) is a pressure-driven process that uses a membrane to remove emulsified oils, metal hydroxides, emulsions, dispersed material, suspended solids and other large molecular weight materials from wastewater, coolant and other solutions. UF excels at the clarification of solutions containing suspended solids, bacteria and high concentrations of macromolecules, including oil and water.

UF systems are designed to reduce oily water volumes by as much as 98 percent without the use of chemical additives. These systems are also capable of removing small fines in deburring and tumbling operations, which allows the water and soap solution to be recycled and reused. When calculating heating and disposal expenses, companies can also see a reduction of wash water and detergent costs by as much as 75 percent and a reduction in waste disposal costs by as much as 90 percent. For these reasons, UF membrane technology is quickly becoming the process of choice over conventional filtration methods.

Vacuum Evaporation and Distillation

Evaporation is a natural phenomenon and a clean separation technology recognized as a best available technique in several wastewater treatment processes. Because it removes the water from the contaminants, rather than filtering the contaminants from the water, it is distinct from other separation processes.

No other technology can attain such high water-recovery and concentration rates as vacuum evaporators, which accelerate the natural evaporation process to treat and distill industrial wastewater amounts from 1 to 120 tons per day. They are capable of achieving residual total solids concentrations of more than 85 percent.

The three primary vacuum evaporators are:

- Heat pumps: Flexible and versatile with low electrical energy consumption and superior reliability.

- Hot water/cold water: Reduce operating costs by utilizing existing excess hot water/steam and cooling water.

- Mechanical vapor recompression: Engineered for the treatment of large wastewater flow rates with low boiling temperatures for reduced energy consumption.

Reverse Osmosis Systems

Reverse osmosis (RO) technology removes dissolved solids and impurities from water by using a semipermeable membrane, which allows the passage of water but leaves the majority of dissolved solids/salts and other contaminants behind. The RO membranes require a greater-than-osmotic pressure and high-pressure water to achieve the desired result. The water that passes through the RO membrane is called the permeate, and the dissolved salts that are rejected by the RO membrane are called the concentrate.

A properly designed and operated RO system can remove up to 99.5 percent of incoming dissolved salts and impurities, as well as virtually all colloidal and suspended matter from the most challenging waste and feedwater applications. Typically for industrial, metalworking and surface treatment applications, RO technology is the final process after UF or the chemical treatment of waste and feedwater.

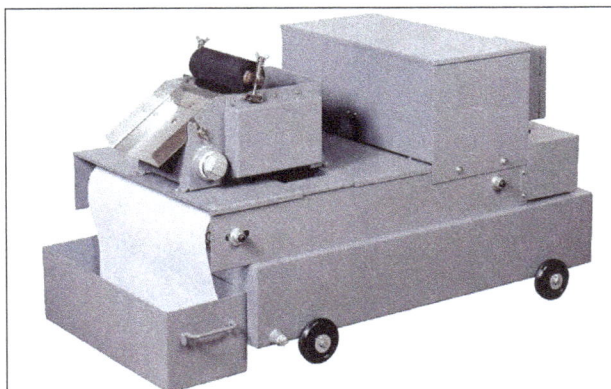

Paper Bed Filters

These types of filters work by gravity and utilize disposable paper media or permanent filter media to produce a positive barrier, which removes solids from all free-flowing industrial process liquids. Paper bed filters are suitable for applications that involve low- to medium-stock removal of ferrous and nonferrous metals, as well as organic and inorganic contaminants such as glass, rubber and plastic. Paper bed filters can extend coolant and tool life by an average of 27 percent, in addition to increasing surface finish quality.

Standard paper bed filtration units are available with or without magnetic separation and can handle flow rates of up to 130 gallons per minute (gpm). Different classes of filter media allow for adjustments of micron clarity. A drum-type model, which can process up to 500 gpm of fluid, occupies one-third the floor space of a paper bed filter.

Solid Bowl Centrifuges

These units optimize centrifugal force (instead of consumable media) to separate solids from liquids in metal processing applications where removal of fines is important for recycle and reuse goals. Process liquid is either pumped or gravity-fed to the centrifuge inlet. Process solids are then centrifugally separated from the liquid phase and collected in an easily removable rotor, also known as a liner. Clarified liquid overflows the rotor into the outer case and is returned by gravity to the process, which minimizes the cost of hauling waste coolants and water away from the facility.

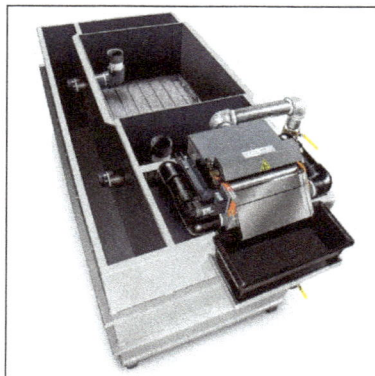

Solid bowl centrifuges provide high-performance liquid/solid separation for all types of particles –
metallic and nonmetallic, ferrous and nonferrous – and are available in both manually cleaned
rotor style (with a reusable liner) and fully automatic self-cleaning designs.

Tramp Oil Separators

In this wastewater treatment solution, contaminated fluid flows through a series of baffles and a
porous media bed, during which free and mechanically dispersed oils are separated from the fluid.
The clarified fluid then flows over the effluent discharge weir back to the reservoir for reuse. The
collected oils, inverted emulsions and other waste materials are collected at the top of the separator and automatically discharged into a suitable receptacle. Using gravity flow and coalescence,
these separators can reduce tramp oils to less than 1 percent in a single pass while utilizing no
consumable products.

Tramp oil separators can also:

- Reduce new fluid purchase costs up to 75 percent (including synthetic and semisynthetic
  coolants, soluble oils and wash waters).

- Reduce the cost of wash water detergents, heating and disposal.

- Reduce hazardous waste volumes up to 90 percent.

- Achieve system payback (or ROI) in six months or less.

Vacuum Filters

Capable of continuous operation, vacuum filtration systems can eliminate significant down-
time. Virtually maintenance-free and delivering high-sludge-volume elimination, these
systems will also deliver lower production costs. Disposable media vacuum filters utilize a vacuum chamber to draw contaminated coolant through the disposable filter media.
By applying the proven principle of optimal filtration through contaminate or sludge build-
up, a filter cake forms on the media. These units are capable of impressive flow rates of up to
2,000 gpm.

Semi-permanent vacuum filters can further reduce operation costs by eliminating the need for disposable media. Back-flushing with clean coolant keeps the filter clean without requiring large volumes of air. The back-flushed solids drop from the filter element and settle into a tank, where they are removed via a chain drag-out/flight arrangement. These units require minimal floor space and are completely self-contained, simplifying maintenance and operation.

Industrial Wastewater Reverse Osmosis

Reverse Osmosis (RO) is a widely accepted unit operation for water purification. The water is typically pressured between 150 to 600 psig and pass through either thin film composite or cellulose acetate membranes. RO water recoveries of 70-90% are typical and salt rejection rates are between 90-99%.

The most important factor in treating industrial wastewater with Reverse Osmosis is the pretreatment that protects the membrane against organic fouling, mineral scaling and chemical degradation. Before reverse osmosis should be considered, a complete cation/anion balance is required and potential foulants must be identified. High BOD and COD levels can also contribute to membrane fouling. A wide range of pretreatment technologies is available.

The effluent discharged to the sewer typically contains between 200 to 10,000 parts per million (ppm) total dissolved solids (TDS). With the proper pretreatment technology followed by RO, this water can be recycled.

Ion exchange treatment of the RO product water can further polish the water and make it suitable for all rinses. To design a successful and cost-effective system, USFilter evaluates each individual application because the pH, oxidizing potential and concentration of soluble salts of the wastewater effluents often exceed the operating limits of the RO systems. After the detailed evaluation of the wastewater is complete, USFilter determines the optimum preconditioning chemistry and selects the best pretreatment technology for the application.

Applications include treatment and recycle of wastewaters generated from metal finishing and plating operations; printed circuit board and semiconductor manufacturing (treatment and recycle of rinse waters used in electroplating processes); automotive manufacturing (treatment and recycle of water used for cleaning and painting); food and beverage (concentration of wastewater for reuse and reduction of BOD prior to discharge); groundwater and landfill leachate (removal of salts and heavy metals prior to discharge).

Features and Benefits

- Can be integrated with an existing membrane filtration system or ion exchange system to achieve up to 80% rinse water recycle.

- Reduces water and sewer use costs.

- Modular design for ease of installation.

Brine Disposal and Treatment

The many options for managing brine, a term for saline wastewater from industrial processes, fall under two categories: brine treatment and brine disposal. Brine treatment involves desalinating the brine for reuse and producing a concentrated brine (lower liquid waste volume), or residual solids (zero liquid discharge). Brine disposal includes discharging brine to sewers, surface water, injection wells, or sending it to environmental service providers.

The cost and environmental impact of each option varies significantly based on many factors. Choosing management options for the waste brine requires careful consideration of applicable discharge regulations, availability of disposal methods, and the economic feasibility to treat the brine.

Brine Water Chemistry

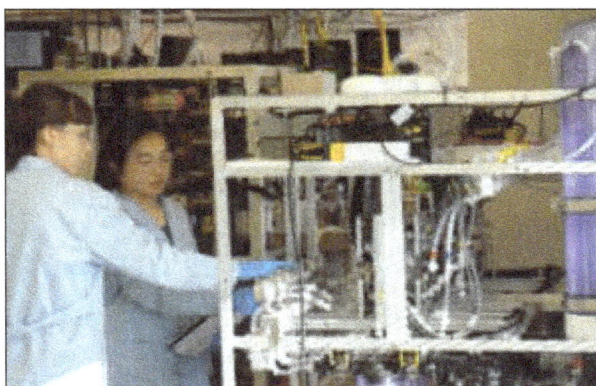

Before deciding on how to manage waste brine, you should consider completing a water chemistry analysis to understand essential indicators, such as the salinity level (e.g. total dissolved solids), metals contaminants, and the scaling potential of the water (e.g. calcium and sulfate). This will assist in evaluating regulatory requirements as well as determining available options and their associated costs.

Water chemistry data provides the most value if you end up deciding to treat your brine. The chemical makeup of the water identifies the technologies that will best fit a specific brine treatment process, for example, whether you should choose thermal or membrane systems. This data enables early assessments of project feasibility and economics, as well as any pretreatment requirements or scaling and fouling risks.

Another advantage of brine chemical characterization is that it allows you to identify opportunities for beneficial resource recovery. For example, it is possible to recover 'fertilizer water' from a waste brine. If a brine contains a mixture of sodium and hardness, electro dialysis reversal (EDR) with certain monovalent ion selective membranes could produce a water high in plant-nourishing hardness with low concentrations of the pollutant sodium. This water would have a low soil adsorption ratio (SAR) that would be valuable to the agricultural industry.

Alternatively, if a brine consists mostly of sodium and chloride, it can be treated with a crystallizer to produce solids that can be used as road de-icing salts.

Brine Treatment

Brine treatment is usually considered if discharge options are not available, brine disposal is expensive, or freshwater recovery is important. There are many technology options to concentrate brine, reduce its volume and disposal costs, or to produce solids for zero liquid discharge. Regardless of the treatment strategy you choose, it will beneficially produce freshwater.

Membrane Treatment Systems

Reverse osmosis (RO) is the membrane system most widely used to desalt brine waters. RO produces freshwater and more concentrated brine often referred to as RO brine, reject, or concentrate. This brine concentrate will usually reach concentrations of dissolved salts and chemicals that will be near scaling limits. This requires treatment to relieve the scaling potential if you will use a thermal system to further concentrate the brine or to produce solids. Alternatively, you could consider thermal systems that can operate under scaling conditions, such as seeded slurry evaporators or a SaltMaker, to eliminate the thermal pretreatment step.

Conventional RO can concentrate brine to a theoretical maximum of 80K mg/L. This is based on the technology's inherent osmotic pressure limit. However, the achievable output brine concentration will usually be less due to the input brine's scaling potential. New Ultra-high Pressure RO (UHP-RO) membranes that operate at up to 1800 psi compared to conventional RO's 1200 psi can reduce brine volume to half that of RO, however scaling must be even more carefully managed due to the higher pressure. Saltworks' XtremeRO is an examples of an RO / UHP-RO systems respectively.

Chemical softening can be used to manage scaling; however, cost, physical footprint, and the ability to deal with varying feedwater chemistry must be considered. Technologies such as Saltworks' automated BrineRefine provide an economic, compact, and flexible chemical softening solution for maximizing RO and UHP-RO brine concentration. Saltworks' BrineGo is an example of a fully integrated solution that combines RO or UHP-RO, BrineRefine, and central control to provide a membrane based solution that concentrates brine to levels only previously attainable by more costly thermal systems.

If your brine contains hydrocarbons or organics, electrodialysis reversal (EDR) may be a better fit than RO due to its lower pretreatment requirements. EDR is a low-pressure system that fluxes salts through ion exchange membranes using an applied electrical charge. There are EDR systems that use anti-fouling ion exchange membranes, such as Saltworks' Flex EDR Organix, that can operate with hydrocarbons and organics present in the brine.

If a thermal system will be used for further concentration or to produce solids then scaling management is still important. Alternatively, you could consider thermal systems that can operate under scaling conditions, such as seeded slurry evaporators or a SaltMaker, to eliminate the thermal pretreatment step.

Thermal Treatment Systems

If you are considering thermal evaporative systems, maximizing freshwater recovery from lower cost membrane systems before using expensive thermal systems will deliver the best project economics. In general, there are two types of thermal systems based on their residual outputs: (1) evaporators that produce concentrated, low volume brine but do not precipitate solids; and (2) crystallizers that exceed salt saturation and produce solids. For high flow rate zero liquid discharge applications, evaporators are used to preconcentrate the brine prior to the crystallizer for final solids production. At lower flows, the waste brine can be sent directly to the crystallizer after treating with a membrane system.

The final disposal of residuals is important in determining whether additional process steps are required. If you have options for disposing of concentrated brine, it will usually not require further treatment. Evaporators are only reducing the volume of brine for final disposal, ensuring you need fewer trucks to move the brine or less capacity in disposal wells or ponds. However, depending on the treatment technology you use, additional treatment may be required for solid residuals before a landfill will accept them for disposal. Almost all landfills require solids to pass a paint filter test, while some also require analysis of pH and leachable metals. To pass a paint filter test, the solids should be dewatered until they have no free water present. Centrifuges, filter presses, and/or dryers are required to further process solids produced by conventional crystallizers to pass the paint filter test. Other crystallizers, such as the SaltMaker, have their own solids management systems that produce dewatered solids in sacks without the need for centrifuges, filter presses, or dryers.

Treatment costs increase the further you concentrate brine towards solids, which is why it is important to carefully consider all disposal and reuse options before implementing a technological solution.

Brine Disposal

Discharging Brine into Surface Bodies of Water or Sewer Systems

If your brine meets regulatory requirements, brine discharge into the nearest body of water or to sanitary sewers is usually the lowest cost option for disposal. Discharge regulations or guidelines vary widely from region to region, or are sometimes determined on a project-specific basis. Regulations may prohibit discharge based on any of the following:

- Concentrations of certain constituents of concern (e.g., maximum limits for metals, salinity, or compounds).

- Total mass per day of certain constituents of concern.

- Specific properties, such as temperature and pH.

- Volumetric flow rates.

- Discharges only during certain time of day.

One option to comply with regulatory discharge requirements may be to dilute the brine stream with other waters requiring discharge. With sufficient dilution, this may reduce the controlled constituents to below the allowable concentration limits. If the brine stream has only one or two constituents of concern that exceed the discharge limits, you should consider selective treatment or removal of those constituents. There are low cost solutions available for removing certain constituents, such as using green sand for iron removal. While discharging brine directly into surface water systems or sewers is often the most cost-effective solution, your organization should consider how it will impact the local environment. If regulations do not exist, studying the potential impacts of discharging the brine on local flora and fauna will help identify the benefits of treatment to protect the ecosystem or prepare for impending regulations.

Brine Disposal in the Ocean

Like discharging brine into surface bodies of water, ocean discharge is another brine disposal method that tends to be very cost effective. In southern California, there is a 'Brine Line' that allows inland plants to discharge their brine to the ocean rather than to sewer or surface waters. Due to the ocean's naturally high salinity, there are lower environmental risks of brine discharge. If you are considering installing a brine discharge line, you will need to acquire a permit. As part of the permit application, the regulatory body may ask for environmental studies that address the impact on local marine ecology of the brine temperature, pH, salt density, and other property differences between the brine and seawater.

Deep Well Injection of Waste Brine

Waste brine can be disposed by injecting it into deep wells. These injection wells are installed thousands of feet deep into the ground, away from the upper aquifers that feed drinking water sources. The availability of injection wells is geology-dependent, so they are not available in all regions. In the oil and gas industry, abandoned oil wells are often converted into disposal wells. Recently, there have been studies that correlate deep disposal wells with increased seismic activity, as evidenced by earthquakes in Oklahoma. Deep well capacities have also reduced as regulations are requiring lower injection pressures to minimize the risk of contaminating the upper water aquifers. Moreover, securing a functioning deep well is similar to drilling for oil – you take a risk and invest capital before knowing if the underground geology will meet your expectations. It is possible that deep wells once drilled will accept very small volumes, or exceed expectations and accept more.

Brine Evaporation Ponds

Evaporation ponds are the artificial solution to inland surface water discharge of waste brine. Under the right climatic conditions, the water evaporates, allowing you to discharge more brine to the ponds. One limitation of ponds is that they require large areas of land to increase the surface area where the water can evaporate, and can represent a future environmental liability due to either animal entry or future decommissioning. If you need to recover solids for disposal or reuse, then multiple evaporation ponds may be necessary to rotate between brine evaporation and solids extraction. Evaporation also happens more quickly in warmer, arid climates. You should consider installing proper liners, preventing waterfowl poisoning from brine that contains metals, and develop an end of life closure plan if your project will be using evaporation ponds.

Brine Incineration

Waste brine can be sent to an incinerator facility, where it is typically mixed with other solid wastes for processing. Incineration evaporates the water, while the salts in the brine become part of the residual ash that requires further management. Incineration is popular in countries with limited availability of land for landfills.

Treatment of Wastewater from Fertiliser Industry

A variety of waste streams are discharged from fertiliser plants in the form of:

- Processing chemicals like sulphuric acid.

- Process intermediates like ammonia, phosphoric acid, etc.

- Final products like urea, ammonium sulphate, ammonium phosphate, etc.

In addition, oil bearing wastes from compressor houses of ammonia and urea plants, and some portion of cooling water and wash water from the scrubbing towers for the purification of gases, also come as waste.

Wash water from the scrubbing towers may contain toxic substances like arsenic, monoethanolamine, potassium carbonate, etc. in a nitrogenous fertiliser plant, while that in a phosphatic fertiliser plant may contain a mixture of carbonic acid, hydrofluoric acid and fluosilicic acid.

Both alkaline and acidic wastes are also expected from the boiler feed water treatment plant, the wastes being generated during the regeneration of anion and cation exchanger units. Waste cooling water contained toxic elements like chromates, zinc, etc. which were used for corrosion control.

The development of non-chromate technology using quaternary ammonium compounds has eliminated these toxic substances. Additional pollutants like phenol and cyanide will be introduced in the list of pollutants in a fertiliser plant where ammonia is derived from the waste ammoniacal liquor of coke ovens.

Effects of Wastes on Receiving Streams

All the components of waste from the fertiliser plants induce adverse effects in streams. Acids and alkalies can destroy normal aquatic life. Arsenic, fluorides, and ammonium salts are found to be toxic to fish. Amines are not only toxic to fish but also exert a high oxygen and chlorine demand.

Presence of different types of salts renders the stream unfit for use as a source of drinking water in the downstream side. Nitrogen and other nutrient content of the waste encourage growth of aquatic plants in the stream.

Treatment of Fertiliser Wastewater

Major pollutants in fertiliser Wastewater for which treatment is necessary include oil, arsenic, ammonia, urea, phosphate and fluoride. Oil is removed in a gravity separator. Arsenic containing waste is segregated and after its concentration, the solid waste is disposed of in a safe place.

Phosphate and fluoride bearing wastes are also segregated and chemically coagulated by lime; clarified effluent, which still contains some amount of phosphate and fluoride, is diluted by mixing with other wastes.

Several alternatives are there for the treatment of ammonia bearing wastes, including:

- Steam stripping.

- Air stripping in towers.

- Lagooning after pH adjustment.

- Biological nitrification and denitrification.

For all practical purposes, 'steam stripping' for ammonia removal from fertiliser wastes has been found to be uneconomical. Removal of ammonia gas from the solution in an air stripping tower, packed with red wood stakes, is found to be a very efficient method.

Very encouraging results are obtained from some laboratory and pilot plant studies conducted by National Environmental Engineering Research Institute (NEERI) in the removal of ammonia by simply lagooning the waste.

It was found that considerable reduction in the ammonia content can be accomplished just by retaining the ammoniacal waste in an earthen tank, about 1 metre deep, for a day or two, after pretreatment of the waste by lime to increase the pH to 11.0.

However, no reduction in urea content was observed within this period in wastes containing urea; thus waste containing both urea and ammonia required to be retained in the lagoon for a longer period, to allow urea to decompose to ammonia first.

Biological nitrification involves oxidation of ammonia to nitrate, via nitrite under aerobic conditions; this is followed by denitrification of the nitrified effluent under anaerobic condition, in which gaseous N2 and N2O are the end products and are released into the atmosphere.

The denitrification requires addition of some quantity of carbonaceous matter in the reactor. In all the ammonia removal methods, urea remains untouched. If urea removal is required, wastes containing urea must be retained for a sufficiently long time in an earthen lagoon to allow it to decompose first to ammonia.

Straight Nitrogenous Fertilisers

Urea and salts of ammonia are referred to as straight fertilisers and if combined with other nutrients such as phosphates and potash, they are called complex/mixed fertilisers. Nitrogenous fertiliser plants use a large quantity of water mainly for process cooling, steam generation and process use, resulting in Wastewater generation at various points in the manufacturing process. In an ammonia plant, if partial oxidation process is used, then the carbon of hydrocarbon feedstock is not completely combusted.

Disposal of the Final Treated Effluent and Monitoring

Disposal of the final treated effluent to the receiving water system is an important aspect in the pollution control system.

The detailed procedure to be adopted for disposal system is given as under:

- It is desired that all the effluents after treatment shall be routed to a properly lined guard pond for equalisation and final control.

- The guard pond should have two compartments, each of at least four hours capacity. All the effluent streams shall be connected to these compartments by a parallel connection system. One compartment of the guard pond shall be used for the routine disposal of effluent, while the other compartment will remain empty and be utilised when effluent do not conform the limits.

- In the guard pond, an automatic monitoring system for flow and the relevant pollutants shall be provided with high-level alarm system. The parameters necessary for automatic monitoring are pH, ammoniacal nitrogen, nitrate nitrogen and hexavalent chromium.

- Monitoring of nitrate nitrogen is applicable to the industries where nitric acid is produced and used for production of fertilisers, and the monitoring of hexavalent chromium is applicable to the industries where chromate-based inhibitors are used for cooling water conditioning.

- When pollutants in the final effluent exceed the stipulated figures as indicated by the alarm system and the effluent stream responsible cannot be identified, all the effluent streams shall flow the empty compartment of the guard pond. The effluent thus stored should be treated before discharge.

- The area around the guard pond shall be developed with proper road connection and lighting system so that it can be approached easily at any time.

- Till the continuous monitoring systems are not installed, the industry shall collect grab samples at a four-hour interval and analyse these for pH and ammoniacal nitrogen.

- The other parameters as relevant to the industry concerned such as total kjeldahl nitrogen, hexavalent chromium, total chromium cyanide, nitrate nitrogen, vanadium, arsenic, suspended solids and oil and grease should be analysed in the grab sample collected once a day at fixed hours.

- For effective appraisal of the performance of treatment units, the industry shall monitor the concerned parameters at least once a shift before and after treatment.

Phosphate Fertiliser

Industrial Operation and Wastewater

The phosphate manufacturing and phosphate fertiliser industry is a basic chemical manufacturing industry, in which essentially both the mixing and chemical reactions of raw materials are involved in production.

Also, short and long-term chemical storage and warehousing, as well as loading/unloading and transportation of chemicals, are involved in the operation. In the case of fertiliser production, only the manufacturing of phosphate fertilisers and mixed and blend fertilisers containing phosphate along with nitrogen and/or potassium is presented here.

Regarding Wastewater generation, volumes resulting from the production of phosphorus are several orders of magnitude greater than the Wastewaters generated in any of the other product categories. Elemental phosphorus is an important Wastewater contaminant common to all segments of the phosphate manufacturing industry, if the phossy water (water containing colloidal phosphorus) is not recycled to the phosphorus production facility for reuse.

The major Wastewater source in the defluorination processes is the wet scrubbing of contaminants from the gaseous effluent streams. However, process conditions normally permit the use of recirculated contaminated water for this service, thereby effectively reducing the discharged Wastewater volume.

Characteristics and Sources of Wastewater from Fertiliser Industry

Wastewaters from the manufacturing, processing and formulation of inorganic chemicals such as phosphorus compounds, phosphates and phosphate fertilisers cannot be exactly characterised.

The wastewater streams are usually expected to contain trace or large concentrations of all raw materials used in the plant; all intermediate compounds produced during manufacture; all final products, coproducts and by-products; and the auxiliary or processing chemicals employed.

It is desirable from the viewpoint of economics that these substances not be lost, but some losses and spills appear unavoidable and some intentional dumping does take place during housecleaning, vessel emptying and preparation operations.

Few fertiliser plants discharge Wastewaters to municipal treatment systems. Most use ponds for the collection and storage of Wastewaters, pH control, chemical treatment and settling of suspended solids.

Whenever available retention pond capacities in the phosphate fertiliser industry are exceeded, the Wastewater overflows are treated and discharged to nearby surface water bodies. The range of wastewater characteristics and concentrations for typical retention ponds used by the phosphate fertiliser industry.

The specific types of Wastewater sources in the phosphate fertiliser industry are:

- Water treatment plant wastes from raw water filtration, clarification, softening and deionisation, which principally consist of only the impurities removed from the raw water (such as carbonates, hydroxides, bicarbonates and silica) plus minor quantities of treatment chemicals.

- Closed-loop cooling tower blowdown, the quality of which varies with the makeup of water impurities and inhibitor chemicals used (the only cooling water contamination from process liquids is through mechanical leaks in heat exchanger equipment. Table highlights the normal range of contaminants that may be found in cooling water blowdown systems).

- Boiler blowdown, which is similar to cooling tower blowdown but the quality differs as given in table.

- Contaminated water or gypsum pond water, which is the impounded and reused water that accumulates sizable concentrations of many cations and anions, but mainly fluorine and phosphorus [concentrations of 8500 mg/l F and in excess of 5000 mg/l P are not unusual; concentrations of radium 226 in recycled gypsum pond water are 60 to 100 picocuries/l and its acidity reaches extremely high levels (pH 1-2)].

- Wastewater from spills and leaks that, when possible, is reintroduced directly to the process or into the contaminated water system.

- Nonpoint source discharges that originate from the dry fertiliser dust covering the general plant area and then dissolving in rain water and snowmelt that become contaminated.

Table: Range of concentrations of contaminants in cooling water.

Cooling water contaminant	Concentration (mg/l)
Chromate	0-250
Sulphate	500-3000
Chloride	35-160
Phosphate	10-50
Zinc	0-30
TDS	500-10000
TSS	0-50
Biocides	0-100

Table: Range of concentrations of contaminants in boiler blowdown.

Boiler blowdown contaminant	Concentration (mg/l)
Phosphate	5-50
Sulphate	0-100
TDS	500-3500
Zinc	0-10
Alkalinity	50-700
Hardness	50-500
Silica (SiO_2)	25-80

In the specific case of Wastewater generated from the condenser water bleedoff in the production of elemental phosphorus from phosphate rock in an electric furnace, Horton reported that the flow varies from 10 to 100 gpm (2.3-23 m³/hr), depending on the particular installation.

The most important contaminants in this waste are elemental phosphorus, which is colloidally dispersed and may ignite if allowed to dry out and fluorine that is also present in the furnace gases. The general characteristics of this type of Wastewater (if no soda ash or ammonia were added to the condenser water) are given in table.

Table: Range of concentrations of contaminants in condenser waste from electric furnace production of phosphorus.

Quality parameter	Concentration or value
pH	1.5-2.0
Temperature	120°-150°F
Elemental phosphorus	400-2500 ppm
Total suspended solids	1000-5000 ppm
Fluorine	500-2000
Silica	300-700 ppm
P2O5	600-900 pp
Reducing substances as (I2)	40-50 ppm
Ionic charge of particles	Predominantly positive (+)

Fertiliser manufacturing may create problems within all environmental media, i.e. air pollution, water pollution and solid wastes disposal difficulties.

In particular, the liquid waste effluents generated from phosphate and mixed and blend fertiliser production streams originate from a variety of sources and may be summarised as follows:

- Ammonia-bearing wastes from ammonia production.

- Ammonium salts such as ammonium phosphate.

- Phosphates and fluoride wastes from phosphate and superphosphate production.

- Acidic spillages from sulphuric acid and phosphoric acid production.

- Spent solutions from the regeneration of ion-exchange units.

- Phosphate, chromate, copper sulphate and zinc wastes from cooling tower blowdown.

- Salts of metals such as iron, copper, manganese, molybdenum and cobalt.

- Sludge discharged from clarifiers and backwash water from sand filters.

- Scrubber wastes from gas purification processes.

Considerable variation, therefore, is observed in quantities and Wastewater characteristics at different plants. The most important factors that contribute to excessive in-plant materials losses and therefore, probable subsequent pollution are the age of the facilities (low efficiency, poor process control), the state of maintenance and repair (especially of control equipment), variations in feedstock and difficulties in adjusting processes to cope and an operational management philosophy such as consideration for pollution control and prevention of materials loss.

Because of process cooling requirements, fertiliser manufacturing facilities may have an overall large water demand, with the Wastewater effluent discharge largely dependent on the extent of in-plant recirculation. Facilities designed on a once-through process cooling flow-stream generally discharge from 1000 to over 10,000 m^3/hr Wastewater effluents that are primarily cooling water.

According to research results reported by Fuller, the removal of semi-colloidal matter in settling areas or ponds seems to be one of the primary problems concerning water pollution control.

The results of Dissolved Oxygen (DO) and BOD surveys indicated that receiving streams were actually improved in this respect by the effluents from phosphate operations. On the other hand, no detrimental effects on fish were found, but there is the possibility of destruction of fish food aquatic micro-organisms and plankton under certain conditions.

The Wastewater characteristics vary from one production facility to the next and even the particular flow magnitude and location of discharge will significantly influence its aquatic environmental impact.

The degree to which a receiving surface water body dilutes a Wastewater effluent at the point of discharge is important, as are the minor contaminants that may occasionally have significant impacts.

Fertiliser manufacturing wastes, in general, affect water quality primarily through the contribution of nitrogen and phosphorus, whose impacts have been extensively documented.

Significant levels of phosphates assist in inducing eutrophication and in many receiving waters they may be more important (growth-limiting agent) than nitrogenous compounds. Under such circumstances, programs to control eutrophication have generally attempted to reduce phosphate concentrations in order to prevent excessive algal and macrophyte growth.

In addition to the above major contaminants, pollution from the discharge of fertiliser manufacturing wastes may be caused by such secondary pollutants as oil and grease, hexavalent chromium, arsenic and fluoride.

As reported by Beg in certain cases, the presence of one or more of these pollutants may have adverse impacts on the quality of a receiving water, due primarily to toxic properties or can be inhibitory to the nitrification process. Finally, oil and grease concentrations may have a significant detrimental effect on the oxygen transfer characteristics of the receiving surface water body.

Wastewater Control Methods for Fertiliser Industry

The pollution control and treatment methods and unit processes used are discussed below:

In-Plant Control, Recycle and Process Modification

The primary consideration for in-plant control of pollutants that enter waste streams through random accidental occurrences, such as leaks, spills and process upsets, is establishing loss prevention and recovery systems.

In the case of fertiliser manufacture, a significant portion of contaminants may be separated at the source from process wastes by dedicated recovery systems, improved plant operations, retention of spilled liquids and the installation of localised interceptors of leaks such as oil drip trays for pumps and compressors.

Also, certain treatment systems installed (i.e. ion-exchange, oil recovery and hydrolyser-stripper systems) may, in effect, be recovery systems for direct or indirect reuse of effluent constituents. Finally, the use of effluent gas scrubbers to improve in-plant operations by preventing gaseous product losses may also prevent the airborne deposition of various pollutants within the general plant area, from where they end up as surface drainage runoff contaminants.

Cooling Water

Cooling water constitutes a major portion of the total in-plant wastes in fertiliser manufacturing and it includes water coming into direct contact with the gases processed (largest percentage) and water that has no such contact.

The latter stream can be readily used in a closed-cycle system, but sometimes the direct contact cooling water is also recycled (after treatment to remove dissolved gases and other contaminants and clarification). By recycling, the amount of these Wastewaters can be reduced by 80 to 90 per cent, with a corresponding reduction in gas content and suspended solids in the wastes discharged to sewers or surface water.

Process Modifications

The following are possible process modifications and plant arrangements that could help reduce wastewater volumes, contaminant quantities and treatment costs:

- In ammonium phosphate production and mixed and blend fertiliser manufacturing, one possibility is the integration of an ammonia process condensate steam stripping column into the condensate boiler feedwater systems of an ammonia plant, with or without stripper bottoms treatment depending on the boiler quality makeup needed.

- Contaminated Wastewater collection systems designed so that common contaminant streams can be segregated and treated in minor quantities for improved efficiencies and reduced treatment costs.

- In ammonium phosphate and mixed and blend fertiliser production, another possibility is to design for a lower-pressure steam level (i.e. 42-62 atm) in the ammonia plant to make process condensate recovery easier and less costly.

- When possible, the installation of air-cooled vapour condensers and heat exchangers would minimise cooling water circulation and subsequent blowdown.

Recently new techniques have been adopted by French company for pollution prevention, for a new process modification for steam segregation and recycle in phosphoric acid production in which, raw water from the sludge/fluorine separation system is recycled to the heat-exchange system of the sulphuric acid dilution unit and the Wastewater used in plaster manufacture.

Furthermore, decanted supernatant from the phosphogypsum deposit pond is recycled for treatment in the water filtration unit. This process modification permits an important reduction in pollution by fluorine and that it makes the treatment of effluents easier and in some cases allows specific recycling.

Finally, the new process produced a small reduction in water consumption, either by recycle or discharging a small volume of polluted process water downstream and required no particular equipment and very few alterations in the mainstream lines of the old process.

Wastewater Treatment Methods for Fertiliser Industry

Phosphate Manufacturing

Nemerow summarised the major characteristics of wastes from phosphate and phosphorus compounds production (i.e. clays, limes and tall oils, low pH, high suspended solids, phosphorus, silica and fluoride) and suggested the major treatment and disposal methods such as lagooning, mechanical clarification, coagulation and settling of refined wastewaters.

Phosphate Fertiliser Production

Contaminated water from the phosphate fertiliser is collected in gypsum ponds and treated for pH adjustment and control of phosphorus and fluorides. Treatment is achieved by 'double liming' or a two- stage neutralisation procedure, in which phosphates and fluorides precipitate.

The first treatment stage provides sufficient neutralisation to raise the pH from 1 to 2 to a pH level of at least 8. The resultant effectiveness of the treatment depends on the point of mixing of lime addition and on the constancy of pH control. Fluosilisic acid reacts with lime and precipitates calcium fluoride in this step of the treatment.

The Wastewater is again treated with a second lime addition to raise the pH level from 8 to at least 9 (where phosphate removal rates of 95 per cent may be achieved), although two-stage dosing to pH 11 may be employed.

Concentrations of phosphorus and fluoride with a magnitude of 6500 and 9000 mg/l, respectively, can be reduced to 5 to 500 mg/l P and 30 to 60 mg/l F. Soluble orthophosphate and lime react to form an insoluble precipitate, calcium hydroxy apatite.

Sludges formed by lime addition to phosphate wastes from phosphate manufacturing or fertiliser production are generally compact and possess good settling and dewatering characteristics and removal rates of 80 to 90 per cent for both phosphate and fluoride may be readily achieved.

The seepage collection of contaminated water from phosphogypsum ponds and reimpoundment is accomplished by the construction of a seepage collection ditch around the perimeter of the diked storage area and the erection of a secondary dike surrounding the first.

The base of these dikes is usually natural soil from the immediate area and these combined earth/gypsum dikes tend to have continuous seepage through them. The seepage collection ditch between the two dikes needs to be of sufficient depth and size to not only collect contaminated water seepage, but also to permit collection of seeping surface runoff from the immediate outer perimeter of the seepage ditch. This is accomplished by the erection of the small secondary dike, which also serves as a backup or reserve dike in the event of a failure of the primary major dike.

The sulphuric acid plant has boiler blowdown and cooling tower blowdown waste streams, which are uncontaminated. However, accidental spills of acid can and do occur and when they do, the spills contaminate the blowdown streams.

Therefore, neutralisation facilities should be supplied for the blowdown waste streams, which involves the installation of a reliable pH or conductivity continuous- monitoring unit on the plant effluent stream. The second part of the system is a retaining area through which noncontaminated effluent normally flows.

The detection and alarm system, when activated, causes a plant shutdown that allows location of the failure and initiation of necessary repairs. Such a system, therefore, provides the continuous protection of natural drainage waters, as well as the means to correct a process disruption.

Mixed fertiliser treatment technology consists of a closed loop contaminated water system, which includes a retention pond to settle suspended solids. The water is then recycled back to the system.

There are no liquid waste streams associated with the blend fertiliser process, except when liquid air scrubbers are used to prevent air pollution. Dry removals of air pollutants prevent a wastewater stream from being formed.

Phosphate and Fluoride Removal

Phosphates may be removed from Wastewaters by the use of chemical precipitation as insoluble calcium phosphate, aluminium phosphate and iron phosphate. The liming process, lime being typically added as a slurry and the system used is designed as either a single or two-stage one.

Polyelectrolytes have been employed in some plants to improve overall settling and clariflocculators or sludge-blanket clarifiers are used in a number of facilities. Alternatively, the dissolved air flotation process is also feasible for phosphate and fluoride removal.

A number of aluminium compounds, such as alum and sodium aluminate, have also been used as phosphate precipitants at an optimum pH range of 5.5 to 6.5, as have iron compounds such as ferrous sulphate, ferric sulphate, ferric chloride and spent pickle liquor.

The optimum pH range for the ferric salts is 4.5 to 5 and for the ferrous salts it is 7 to 8, although both aluminium and iron salts have a tendency to form hydroxyl and phosphate complexes. As reported by Ghokas, sludge solids produced by aluminium and iron salts precipitation of phosphates are generally less settleable and more voluminous than those produced by lime treatment.

According to Sprecht, in the two-step process to remove fluorides and phosphoric acid, water entering the first step may contain about 1700 mg/L F and 5000 mg/L P_2O_5 and it is treated with lime slurry or ground limestone to a pH of 3.2 to 3.8.

Insoluble calcium fluorides settle out and the fluoride concentration is lowered to about 50 mg/l F, whereas the P_2O_5 content is reduced only slightly. The clarified supernatant is transferred to another collection area where lime slurry is added to bring the solution to pH 7 and the resultant precipitate of P is removed by settling.

The final clear water, which contains only 3 to 5 mg/L F and practically no P_2O_5, is either returned to the plant for reuse or discharged to surface waters. The two- step process is required to reduce fluorides in the water below 25 mg/l F, because a single-step treatment to pH 7 lowers the fluoride content only to 25 to 40 mg/l F.

In the process where the triple phosphate is to be granulated or nodulised, the material is transferred directly from the reaction mixer to a rotary dryer and the fluorides in the dryer gases are scrubbed with water. In making defluorinated phosphate by heating phosphate rock, one method of fluoride recovery consists of absorption in a tower of lump limestone at temperatures above the dewpoint of the stack gas, where the reaction product separates from the limestone lumps in the form of fines.

A second method of recovery consists of passing the gases through a series of water sprays in three separate spray chambers, of which the first one is used primarily as a cooling chamber for the hot exit gases of the furnace. In the second chamber, the acidic water is recycled to bring its concentration to about 5 per cent equivalence of hydrofluoric acid in the effluent, by withdrawing acid and adding freshwater to the system.

In the final chamber, scrubbing is supplemented by adding finely ground limestone blown into the chamber with the entering gases. Hydrochloric acid is sometimes formed as a by-product from the fluoride recovery in the spray chambers and this is neutralised with NaOH and lime slurry before being transferred to settling areas.

Treatment of Wastewater from Iron and Steel Industry

The iron and steel industry, includes pig iron production, steel making, rolling operations and those finishing operations common in steel mills, i.e. cold reduction, tin plating and galvanising. Most steel firms operate iron ore mines, ore beneficiation plants, coal mines, coal cleaning plants and coke plants; many have fabricating plants or produce a variety of speciality steel products.

Manufacturing Operations

Manufacturing operations of the iron and steel industry may be grouped as pig iron manufacture, steelmaking processes, rolling mill operations and finishing operations. A single mill is not likely to incorporate all of the many combinations and variations of these operations that are possible. Most mills specialise in the production of broad categories of steel products; in a large mill, however, the product list is long.

The manufacture of pig iron is accomplished in the blast furnace. Steel-making processes include pneumatic processes, open hearth processes and electric furnace processes. Rolling mill operations include rolling of blooms, slabs and billets; scarfing and other preparations of semi-finished steel; rolling of shapes, bars, strip and plates; wire drawing; tube drawing and pipe forming; and pickling or other oxide removal operations. Finishing operations include tin plating, galvanising, cold reduction and coating.

Blast Furnaces

The blast furnace process consists essentially of charging iron ore, limestone and coke into the top of the furnace and blowing heated air into the bottom. Combustion of the coke provides the heat necessary to attain the temperatures at which the metallurgical reducing reactions take place.

The incandescent carbon of the coke accounts for about 20 per cent of the reduction of the iron oxides; the carbon monoxide formed between the coke and the oxygen of the blast accounts for the remaining reduction accomplished. The function of the limestone is to form a slag, fluid at the furnace temperature, which combines with unwanted impurities in the ore.

Two tons of ore, 1 ton of coke, ½ ton of limestone and 3½ tons of air produce approximately 1 tonne of iron, ½ ton of slag and 5 tons of blast furnace gas containing the fines of the burden carried out by the blast; these fines are referred to as flue dust.

Characteristics of Steel Mill Wastes

Wastes from the various operations in steelmaking vary widely in characteristics and in volume water pollutant in a typical steel mill complex. These wastes generally have physical and chemical effects on receiving streams different from the oxygen-consuming characteristics of municipal sewage and organic industrial wastes.

Because the waste streams vary so widely and are usually separated by the distances between the several operations, composite effects are of little significance; treatment and disposal generally must be considered for the separate wastes.

Water use in the Industry

The water requirements of steel plants vary widely, depending primarily upon the quantity and quality of the available supply. The use of as little as 1500 gal of water per ton of product has received much attention in one instance where recirculation is extensively practiced, due primarily to short supply.

A figure of 65,000 gal per ton of product has also been widely quoted and has been valid in certain installations that have had practically unlimited water supply. The use of 30,000-40,000 gal of water per ton of product has been typical of many large plants; actual consumptive use of water, i.e. water withdrawn but not returned, is probably less than 1000 gal per ton of product.

A recent industry survey indicated a maximum water use of 49,000 gal per ton of product and an average use of 17,000 gal per ton. Water use in the various departments of a typical integrated mill is approximately as shown in Table. Most of the water required by a steel plant is used for indirect cooling and needs no treatment, provided it is not excessively hard; chlorination is often desirable to prevent slime formation.

Table: Water use in an integrated steel mill.

Volume		
Department	Gallon per ton of finished steel	Per cent of total
Blast furnace	10000	25
Open hearths	5000	12½
Coke plants	5000	12½

Hot mills	10000	25
Pinishing mills	8000	20
Sanitary, boiler and other uses	2000	5
Total	40000	100

The water used in Blast furnace gas washing and in hot mills for roll cooling and scale transport is not necessarily of high quality; it is usually used as pumped. In the various finishing operations such as cold reduction, stainless strip rolling, electrolytic tin lines and galvanising, purer water is required and treated water is often used.

Wastewaters

The various Wastewaters from a typical steel plant are considered here individually, roughly segregated according to the operations from which they result. It must be remembered, however descriptions of the various operations that no such clearcut segregation exists in actual practice. Indeed one of the major problems in installing waste treatment facilities in older mills is the segregation of waste streams from integrated operations.

Gas-washer Waters

The water used in washing Blast furnace flue gas contains from 1000 to 10,000 mg/l of suspended solids, depending upon the furnace burden, size of the furnace, operating methods employed and type of gas-washing equipment.

Following a 'slip' in the furnace, the concentration of solids in washer water may exceed 30,000 mg/L. The use of fine ore and high blast rates result in the highest concentrations of solids in the washer water; the top pressure used in the furnace is also an important factor.

The efficiencies of dry dust catchers and wet washers vary considerably and account for many of the differences found in various installations. Conventional wet washers use an average of about 3000 gpm of water; the newer venturi scrubbers use 600-1000.

The wash water from electrostatic precipitators adds little to the washer flow, but increases the concentration of the finest particles in the waste stream. Blast furnaces producing ferromanganese have a high percentage of semi-colloidal dust particles in the washer water.

The fume from pneumatic steelmaking processes, open hearth and electric furnaces and hot scarfing operations is often eliminated by electrostatic precipitators or venturi scrubbers which produce waterborne wastes. These suspensions are generally similar to gas-washer water, but the particles are much finer.

The flue dust particles in washer water are probably 50 per cent finer than 10 microns and approximate the composition of the furnace burden; the specific gravity is about 3.5 on the average. Effects on the receiving streams include objectionable colour and interference with aquatic life through formation of bottom deposits and impedance of light transmission; in extreme cases sludge banks are formed that interfere with navigation.

Gas-washer waters, especially from furnaces operating on ferromanganese, contain appreciable though highly variable concentrations of complex cyanides and may have a toxic effect on aquatic life.

Scale-bearing Waters

Scale-bearing water originates in the various rolling mill operations and consists of the water used to dislodge scale and to cool the rolled product, plus the water used to transport scale through the flumes beneath the mill line.

The characteristics and quantities of scale-bearing water vary widely depending upon the particular rolling operations. The total iron loss in the form of scale averages about 2½ per cent, from the blooming mill through the final rolling operation.

Scale produced in the rolling of blooms and slabs in primary mills is relatively coarse material and most of it settles out of suspension readily. The scale particles from such mills are 90 per cent or more coarser than 200 mesh. Scale produced in a billet mill is considerably finer; 25 per cent of such particles may be finer than 200 mesh.

The water use in primary mills ranges from 2000 to 7000 gpm, depending upon the design of the mill and the rolling practices. The scale particles are mixtures of various iron oxides, with the higher oxides pre-dominating in scale from primary mills.

The specific gravity of mill scale is about 5.0, hence such particles, particularly the coarse material, tend to clog sewers and to deposit in receiving streams. These are, in general, the only effects of primary mill flume water.

The scale produced in finishing mill rolling operations has, in general, the same composition and specific gravity as that from primary mills. It differs, however, in particle size and quantity and in its effects. Considerably greater variation occurs among installations than in the case of primary mills.

10 to 20 per cent of the scale particles from finishing mills is smaller than 200 mesh. The coarse particles are still relatively fine and the finest particles may be 5 microns or less in diameter. Water use in finishing mills ranges from 5000 gpm or less in bar mills and cold reduction mills to 25,000 gpm or more in the newest hot strip mills.

Finishing mill flume water may settle in the receiving streams and form bottom deposits or sludge banks and may increase the turbidity of the stream or impart an objectionable colour if the scale particles are extremely fine.

Acid Waters

Spent pickling solutions and acid rinse waters differ widely in quantity, composition and concentration, depending upon the manner of pickling, production rate, type of steel being cleaned and the degree to which control over the operation is practiced.

Spent pickling solutions of various types may be produced in different, separated operations at a large mill. Acid rinse waters have the same relative proportions of iron salts and free acid as pickling solutions, but are much more dilute; 10-15 per cent of the acid used in pickling is discharged in rinse waters.

Spent pickling, solution discharges may inhibit bio-oxidation processes in streams and may be injurious to aquatic life if the quantity released in relation to the stream flow is sufficient to lower the pH of the stream significantly.

Sulphuric acid comprises about 90 per cent of the total of acids used in pickling steel. Spent sulphuric acid pickling solutions contain free acid, ferrous sulphate, undissolved scale and dirt and the various inhibitors and wetting agents, as well as dissolved trace metals.

The spent solutions from continuous strip picklers contain 5-9 per cent free acid and 13-16 per cent ferrous sulphate; from batch operations the spent solutions may contain 0.5-2.0 per cent free acid and 15-22 per cent ferrous sulphate.

10 to 15 per cent of the acid used in pickling is discharged in the rinse water as highly diluted free and combined acid. Spent sulphate pickling solutions are discharged at 170°F-190°F and can amount to 1,00,000 gal per day in a large mill.

Hydrochloric, nitric, phosphoric and hydrofluoric acids are used in pickling stainless steels. These acids may be used alone, in various combinations, or in combination with sulphuric acid. Stainless steel pickling practices vary widely, in the industry.

A typical pickling operation for stainless steel plates consists of a 10 per cent sulphuric acid bath at 160°F, followed by a 10 per cent nitric acid, 4 per cent hydrofluoric acid bath at 150°F. A typical continuous pickling line for stainless steel strip consists of a 15 per cent hydrochloric acid bath at 160°F followed by a 4 per cent hydrofluoric acid, 10 per cent nitric acid bath at 170°F.

Phosphoric acid is often used in pickling when a phosphate coating is desired. Hydrochloric acid is being used increasingly for mild steels in the new vertical tower pickling installations. These spent solutions contain free acids and the various iron salts, as well as undissolved scale and dirt, inhibitors, wetting agents and trace metals. Compositions vary widely according to the specific operation and plant practice.

Oil-bearing Waters

Rolling oils, lubricants and hydraulic oils are present in the effluents from many operations in a steel mill and occur as both free and emulsified oils. Volumes of the waste streams and the concentrations of oil vary widely according to operating practices and housekeeping methods.

Emulsified oil in an effluent can be esthetically objectionable and may add a significant BOD. Free oil is particularly objectionable in a stream because very small quantities can result in widespread surface films; larger quantities foul boats and docks and result in unsightly accumulations along stream banks. Severe oil pollution can have serious adverse effects on aquatic life, birds and land animals.

So-called soluble oils are present in the waste discharges from cold reduction mills, electrolytic tin lines and a variety of machine shop operations. Natural palm oil and synthetic proprietary substitutes are used in these operations and form stable emulsions when mixed with water at elevated temperatures, especially when kerosene and various detergent cleaning compounds are used.

Concentrations of soluble oils vary according to the degree of recirculation practices; volumes of the waste streams likewise vary widely. Typically, the effluent from once-through use in a cold reduction mill will contain 200 mg/L oil, 25 per cent of which is a stable emulsion.

Lubricating oils and hydraulic fluids are present in the effluents from all rolling operations and most other machine operations. These oils exist mostly as free, floating films and the quantities depend primarily upon machine maintenance and manual lubrication practices, i.e. upon housekeeping.

Other Wastewaters from steel mills include alkaline cleaning solutions, water used in granulating slag and cooling water. Alkaline cleaning solutions are used to remove rolling oils prior to finishing operations. Caustic soda, soda ash, silicates and phosphates are common cleaning agents.

Spent cleaning solutions contain saponified oils and dirt and have substantial residual alkalinity. Total volumes are small, ranging from 1000 to 10000 gal per week for individual operations; these quantities are usually dumped batchwise. The effects on the receiving stream are probably not adverse, especially if the volumes of acid wastes are relatively large, as is usual.

The quenching of blast furnace slag produces small quantities of water containing slag particles. Effluent from a slag pit may range from 100 to 200 gpm and is usually of a clear appearance. The highly abrasive nature of the suspended slag particles is the principal objection to such effluents; the bulk gravity of slag particles ranges from 0.8 to 1.5 because of expansion in the granulating process.

Cooling water discharges comprise the largest percentage of steel mill effluents and are usually 10°-15°F warmer than the water withdrawn from the source of supply. The rise in temperature is the only change in water used for indirect cooling and is usually not significant if the effluent is discharged into a reasonably large stream.

Discharged cooling water can have an adverse effect at certain plants where the receiving stream is small and supports a temperature-sensitive fish population. More often than not, the cooling water discharge is of better quality than the water withdrawn from the stream, because of the treatment used for corrosion control.

Disposal of Steel Mill Wastes

Methods of waste disposal in the steel industry vary widely from plant to plant. The age of the mill is probably the most important single factor accounting for these variations. In older mills, space for large treatment facilities is often not available; the space required may well be of more potential value for production facilities than the total direct costs of a waste treatment plant.

Other factors influencing the variability in methods include the effluent standards that are applicable, the attitudes of management and the competitive position of the particular operations involved, as well as the characteristics of the waste streams.

Gas-washer Water

Blast furnace gas-washer water is usually treated by plain gravity sedimentation in mechanically cleaned circular clarifiers or in simple rectangular sedimentation basins. Circular clarifiers are used almost exclusively in newer installations.

A typical modern installation may consist of a 75 ft diameter clarifier, handling gas-washer water at 6.9 mgd from a blast furnace rated at 1200 tons per day. The effluent would contain approximately 80 mg/L suspended solids; the underflow of 180 tons per day of wet dust would be pumped to the sinter plant for additional thickening and filtration on leaf type vacuum filters.

Older installations might be typified by two 13 ft x 111 ft rectangular sedimentation basins handling gas-washer water at 3.5 mgd from a Blast furnace rated at 880 tons per day. The effluent concentration might average 250 mg/L suspended solids. Sludge would be dredged from the basins by clamshell buckets at the rate of about 22 tons of wet dust per day and hauled to the sinter plant in railroad cars.

The clarified gas-washer water may be recirculated either wholly or in part. Recirculation is practiced when supply conditions dictate water economy and usually requires secondary treatment such as chemical flocculation. The effluent from simple rectangular basins is not ordinarily suitable for recirculation without such extra treatment.

The use of separate clarifiers for each Blast furnace or pair of furnaces may result in the discharge of untreated wastes whenever clarifier operation is interrupted. A more satisfactory arrangement consists of collecting the gas-washer water from all Blast furnaces, with treatment in centrally located clarifiers. Interconnection of the clarifiers insures continuous treatment even if the operation of one clarifier is interrupted for an extended period.

Some plants operate the washer water clarifiers in series, the underflow of each being added to the influent of the next. A single line to the sinter plant and the agglomerating effect of added sludge are possible benefits of this scheme.

Where effluent requirements are stringent or where existing equipment is called upon to handle greater than design flows, chemical flocculation or the various polyelectrolytes may be used, usually in secondary treatment units.

Polyelectrolytes alone have usually not resulted, in plant practice, in the rather spectacular improvements indicated by laboratory experiments. With improved methods of determining and controlling optimum dosages and with probable price reductions, these materials will doubtless become more commonly used.

The design of circular clarifiers and rectangular sedimentation basins for gas-washer water requires specialised techniques; the conventional criteria for sanitary wastes are not satisfactory. The methods outlined here on scale-bearing waters are generally applicable for gas-washer water clarification.

Scale-bearing Waters

Rolling mill flume water has long been partially clarified in small, simple sedimentation basins known as scale pits, in order to prevent sewer clogging. These pits are usually small in relation to the water flow and the deposited scale is cleaned out periodically with clamshell buckets. In newer mills, scale pits are larger and are designed with the objective of water pollution control; continuous mechanical cleaning is often incorporated.

Flume Water Clarification

A scale pit typical of older practice was 18 ft wide, 30 ft long and 8 ft deep, to handle flume water at the rate of 3500 gpm from the slab rolling section of a hot strip mill. Effluent concentration averaged 200 mg/l of suspended solids; there was no provision for removing oil. The pit effluent went directly to a river. Flume water from the finishing end of the mill contained only line scale, not likely to clog the sewer and went to the river untreated.

When the mill cited above was rebuilt, flume water treatment was improved to provide more effective pollution control. The scale pit was tripled in size and handles all water from the mill; oil is removed continuously through split-pipe skimmers.

The scale pit effluent goes to a 35 ft diameter clarifier for additional solids removal and oil separation and the clarifier effluent is returned to the mill for reuse. Little or no Wastewater from this mill is now discharged.

Treatment following once-through use in newer mills usually consists of primary clarification in a scale pit and secondary clarification and oil removal in relatively larger rectangular sedimentation basins.

Scale pits are typically cleaned by dredging with clamshell buckets and secondary clarifiers are usually cleaned continuously by scrapers on endless chains. Many variations of this basic scheme are found in various rolling mills; in fact, few installations are identical.

Often the secondary clarification includes chemical treatment, typically with additions of lime, ferric sulphate and polyelectrolyte coagulating agents. Chemical treatment is most often used with circular clarifiers as the secondary basins.

When chemical treatment is used, the water is generally reused in the mill; it may be passed through cooling towers, especially in the warm weather months. The recovered scale is sintered for use in the Blast furnace or open hearth furnaces.

Mill scale is comparable to high grade iron ore and is thus a salvaged material of considerable value. Generally speaking, the recovery of mill scale from primary scale pits shows an economic return; more than 90 per cent of recovered scale is obtained from these pits. Scale removal from secondary pits must be justified on the basis of pollution control or as necessary for water reuse.

Steel Mill Wastewater Sedimentation

Research sponsored by the American Iron and Steel Institute has resulted in design procedures that are, applicable for steel mill sedimentation equipment, including scale pits, secondary basins for scale removal and gas-washer water clarifiers.

These procedures predict basin performance in terms of an empirical measure known as the Sedimentation Index (SI), expressed in minutes. The sedimentation index may be interpreted as the settling time, under specified laboratory conditions, that will result in sedimentation equal to that of a particular basin at a specified flow rate.

Values of SI are approximately 0.10 for simple scale pits, 1.0 for secondary mill scale basins, 10.0 for small gas-washer water clarifiers and 30.0 for large washer water clarifiers. This work has shown that there are optimum ratios of width and depth to length for rectangular basins and that large circular clarifiers have less volumetric efficiency than smaller clarifiers at comparable flow rates. Overflow rates and superficial linear velocities are not adequate criteria for the design of steel mill sedimentation equipment.

Rolling mill flume water and Blast furnace gas-washer water should be sampled with care when such samples are to be used as the basis of basin design for required effluent concentrations. Composite samples should be taken over periods of typical operation; samples should be randomised so as not to coincide with process cycles such as slab rollings or Blast furnace chargings.

Settling rate tests should be made soon after collection because many of these suspensions cannot be effectively reconstituted after settling has occurred. Existing installations similar to contemplated new installations can often be used as sample sources for design purposes, but differences in raw water quality due to location and season of the year should be borne in mind.

Pickling Solutions

Few industrial wastes have received as much attention as spent pickling solutions in terms of research and process development effort. The recovery of by-products from waste treatment processes seems attractive, but has not proved economically sound. Relatively dilute solutions of cheap bulk chemicals are involved and the quantities are large in comparison with most possible markets for by-products.

Other Wastewaters

Other steel mill Wastewaters such as alkaline cleaning solutions, slag pit effluents and sanitary wastes present few special problems, but are important in planning pollution control comprehensively.

Alkaline cleaning solutions may be used as additional alkaline agents in spent pickling solution neutralisation, or may simply be diluted prior to discharge if the quantities are relatively small. Slag pit effluents are treated by rotary screening if discharge is to a navigable stream or a recreational stretch of a stream.

Sanitary wastes may be conventionally treated in a mill-operated facility or sent to a municipal sewage treatment plant. The greatest problem encountered with sanitary wastes is in segregating them from process waste streams, especially in older mills; the cost of sewer segregation is usually the greatest cost of treating these wastes.

Water Reuse

Reuse of water in the steel industry will increase in the future. Some of this increased reuse will be for the purpose of conservation in localised situations of water shortages, as in circumstances where the low flow period reduces surface streams to a critical point or where the groundwater faces serious depletion.

The principal factor influencing reuse will probably be the increasing requirement for high effluent quality and the criterion for the extent of reuse will be economics. The completely closed process water system is, of course, the final answer in industrial Wastewater treatment.

Even under conditions of abundant supply, complete recirculation can become economical when effluent quality requirements become sufficiently high. Such system will probably provide the solution to Wastewater control problems in the steel industry increasingly in the future.

Nonferrous Metals

Current usage divides all metals into three groups: iron and steel including alloy steels; ferroalloys; and nonferrous metals.

The subject matter is confined to the processes and related operations intermediate between production of metal ores and the finished product. These intermediate processes are usually those involved in the extraction or refining of commercially pure metal from ores and the fabrication of the metal into usable shapes.

The four major nonferrous metals are aluminium, copper, zinc and lead.

The processes described are concerned with either the extraction of pure metal from the ore or fabrication of the metal. Extraction of pure metal includes a variety of purification methods, such as dissolving metal compounds by leaching, production of oxides, reduction of the oxides to metal by smelting and refining by electrolysis.

Smelters and refiners are primary or secondary, depending on whether they use natural ores or scrap as their principal source for metals. Fabrication of metals includes such operations as alloying, casting, extrusion, forging, rolling, wire-drawing and heat treating and provides sheets, wire, tubing and other industrial shapes.

Major Nonferrous Metals

Production of aluminium from bauxite ore includes an aqueous extraction of aluminium oxide (alumina), followed by electrolytic reduction of molten alumina. Almost 50 per cent of the aluminium produced is made into sheets, including plates and foil; about 25 per cent into extruded tubing; and about 20 per cent into castings. Much of the rest is made into rolled shapes.

Copper is extracted from sulphide concentrates and from 'cement' copper by smelting in a reverberatory furnace to produce anodes of about 98 per cent copper, followed by electrolytic refining using aqueous sulphate solutions.

Most of the ores, used are sulphides. Copper oxides are converted to 'cement' copper by leaching the ores with sulphuric acid and precipitating the copper with scrap iron. About 55 per cent of the copper is fabricated in wire mills and about 40 per cent in brass mills.

In the rolling and drawing of tubes and wires in several steps, the metal tends to become hard and annealing is required after every two or three steps. The oxide scale formed in annealing is removed by dipping the metal products in sulphuric acid baths, followed by rinsing in water.

The major Wastewaters in the copper and brass industry are these rinses; they contain a considerable amount of dissolved copper, zinc, chromium and sulphuric acid. The acid or 'pickle' baths, although they are dumped only infrequently, provide Wastewaters containing the same toxic compounds; spent liquor wastes may be considered related to rinse-water wastes, but are of higher concentration and much lower volume.

Oil-bearing Wastewaters are formed from lubrication, similar to those formed in the aluminium industry. These are frequently discharged into municipal sewers or rivers with little or no treatment.

Characteristics and disposal of this type of waste are discussed under aluminium. Other wastes of the copper and brass industry are the solid scrap, almost all of which is recovered for reuse and zinc fume from the electrolytic melting furnaces, most of which is discharged in stack gases without treatment.

Pickle Rinse Waters from Fabrication

Rinse water and acid bath dumps are discussed together because both contain the same noxious compounds—copper, zinc, chromium and acid—and are related in other ways. Of these two wastes, rinse waters contain the larger mass of contaminant—90 per cent of the total in one study. Although acid bath dumps are more concentrated, the flow rates of the relatively dilute rinse waters are large, averaging 200-1000 gal/ton of product.

Rinse water concentrations vary with time and with the individual plant, but some not at typical values are indicative. Pickle baths are batch vessels of about 1000 gal capacity, filled with a 5-10 per cent sulphuric acid solution.

During the time they are used for pickling, the acid content becomes depleted and the metal content accumulates. When spent, the pickle liquor is discarded and a new batch of acid is prepared. Dumping cycles vary, but are frequently once a month.

Bright dip baths, used to remove stains on the finished tube or wire, operate similarly to pickle baths except that 3-8 per cent sodium dichromate is added to fresh batches and dumping cycles are usually every week or every few days.

Typical compositions of pickle baths contain in mg/l: 80,000 sulphuric acid, 10,000 copper and 10,000 zinc, with maximum values 2-4 times these concentrations. Bright dip baths, when dumped, have similar copper content, somewhat lower acid and zinc content and substantial chromium content (20,000 mg/L).

Discharge of these wastes without treatment is toxic to aquatic life and harmful to sewers and sew age plants; dilution is seldom adequate. Acidity of water below a pH of 6 is often lethal to the aquatic life that forms food for fish. The presence of copper, zinc or chromium above 2 mg/l is lethal to fish; furthermore, natural purification of a stream is inhibited.

In a study, the lethal concentration for salmon was reported as 0.05 mg/l copper or 0.6 mg/l zinc. Permissive metal concentrations are usually set between 0.02 and 1.0 mg/l. These metal sulphate wastes are acidic and corrosive, so they reduce the life of municipal sewers and corrode sewage treatment plant equipment. These wastes also interfere somewhat with the biological treatment of sewage and are not completely removed by municipal sewage treatment processes.

Pickle wash waters from the copper industry have most of the undesirable qualities possessed by steel industry pickle wastes and copper, zinc and chromium are considerably more toxic than iron. The most objectionable liquid wastes in the production and fabrication of copper are the iron sulphate solutions from leaching of oxide ores and the pickle rinse waters from wire and brass mills. Another, but minor, Wastewater is the process water effluent containing entrained solid.

Zinc and Lead

Zinc and lead wastes are similar to wastes from the copper industry; the major ores are sulphides and the primary production of metal is by smelting and electrolysis, except that most of the zinc is refined by distillation. Ores of zinc and lead often occur in the same deposit, sometimes with copper deposits.

Other Metals

Magnesium

In the production of magnesium hydroxide, the water effluent contains fewer salts than the influent sea water used as raw material. Furthermore, magnesium salts are not toxic in the usual concentrations found in water. Little Wastewater is formed during production of the metal by electrolysis; recovery of the metal and of the chlorine gas evolved is almost complete.

Gold

Gold is extracted from gold ores by cyanidation, amalgamation and rest from placers or base metal ores, but only which produces a potential Wastewater hazard. This hazard is due to the high toxicity of cyanide and arises not only from the sodium and calcium cyanides used to form complexes with gold, but also from the cyanide sometimes used in the flotation circuits of mineral beneficiation plants.

After the gold cyanide complexes are split or electrolysed to remove the gold, the cyanides are frequently recycled, thus minimising waste discharge. Waste cyanide solutions can be oxidised to nitrogen and carbon dioxide for electroplating wastes.

Thus, the major potentially objectionable Wastewaters arising from primary production and fabrication of nonferrous metals are the mud slurries from bauxite, fluorine solutions from aluminium refining, oil- bearing wastes from lubrication in fabrication, iron sulphate solutions from copper oxide leaching, pickle washings from brass mills, entrained solids from many operations and cyanides from gold extraction. Of all these, pickle wash waters from brass mills are probably the most objectionable in quantity and toxicity.

Waste Disposal

From a long-range view, the best solution to waste problems is waste elimination by process changes. Wastes may be eliminated in existing plants by reuse of the noxious material or by recovery of by-products for sale; in new plants, choice of an alternate process may eliminate production of the undesirable material.

Examples of these approaches in the nonferrous metal industries have been mentioned. Even where these approaches are not used, process changes to minimise waste formation have reduced treatment costs and decreased the amount of waste discharged to the surroundings. Treatment and disposal should be considered only where process alternatives are not available or where a temporary expedient is needed.

Where wastes are noxious materials that can be converted to harmless form by chemical reactions, treatment may involve such decomposition. In the nonferrous industries, however, noxious materials usually persist through treatment and ultimate disposal becomes significant.

Treatment of such waste- waters often involves separation of the deleterious substances from water by chemical reaction and phase change to gas or solid form, usually designed so that more concentrated mixtures of the noxious material are produced.

Less frequently, treatment may involve concentration to another liquid phase; examples of these concentration treatment methods include ion-exchange and the separation of immiscible liquids. On the other hand, disposal of these materials involves the permanent or semi-permanent relocation of the waste.

Frequently called 'ultimate disposal', this includes dispersion of wastes by dilution and storage of solid materials or slurries on land. Therefore, for these persistent materials, treatment in itself is not adequate, but must be followed by some form of disposal.

Waste disposal may be accomplished, in one or more steps, with or without treatment—either of which may be satisfactory or unsatisfactory, depending upon such conditions as nature of the surroundings and concentration and amount of waste.

An example of unsatisfactory treatment, indicative of some parts of the nonferrous metal industry today, occurs when a gaseous waste is converted into a noxious liquid waste, as in the scrubbing of fluoride gases in the aluminium industry.

This treatment is unsatisfactory at many locations because it is incomplete; satisfactory disposal in these cases requires subsequent treatment and disposal of the new form of the waste. Before considering disposal practices for specific wastes, three aspects common to most of the nonferrous metal industry should be examined: effect of surrounding, plant size and waste concentration.

The primary production of these metals often occurs in arid, sparsely populated areas, whereas much of the secondary production and fabrication is located in or near heavily populated industrial centers where water is often more plentiful.

Much of the primary aluminium, however, is produced in sparsely settled areas having plentiful water. A popular method of treatment and disposal in sparsely populated areas is lagooning, because of low land values. This method, however, is often considered too expensive within heavily populated industrial areas, even though smaller plants are the rule.

Partly because of the more favourable attitude on the part of large companies, which usually operate larger facilities, it is generally true that treatment of wastes from primary production is better accepted than adequate treatment for wastes at small fabrication plants.

It also is generally true that waste disposal in these industries is more of a problem in industrial centers where water is plentiful, but where treatment costs are higher and the accumulation of wastes from several industries is more likely to occur.

These, of course, are generalisations for which notable exceptions occur and which will probably be changed in time. Wastewaters contaminated by metal ions in large concentrations offer the best possibility of economic extraction of the metals for reuse or sale as by-products; treatment costs are often low, as in precipitation of iron sulphate by evaporation in the copper industry.

Dilute solutions, because of their large water volume, are so costly to treat that the practice of discharge of these solutions to large waterways is widespread. This dilution method is possible only if sufficiently abundant water flows are near and only as long as it is condoned by the public and the government.

Where valuable metals are involved, as in electrolytic copper wastes, it is usually economical to concentrate the solutions and to recover the metals. If dilute solutions cannot be stored and re-used, the alternative is expensive treatment; dilute wastes arising from the washing of pickled metal are a current example of this problem. Other rinse waters and cooling waters used in fabrication also pose the problems of dilute solutions.

The treatment method most frequently used in the nonferrous metal industries is sedimentation of solids; it is used for entrained particulate matter as well as for precipitates formed by evaporation and by treatment with alkaline chemicals. Disposal methods most frequently used for Wastewaters of this industry are discharged into rivers and oceans and discharge of solids onto land areas.

Copper

Iron Sulphate Solutions from Leaching

Leaching wastes are formed in the iron launders used to extract copper from oxide ore and to recover copper from tailings of sulphide ore, low grade ore and mine waters. Because these operations are located mainly in the arid rocky area, treatment of this waste is necessary to prevent making streams unpotable and unfit for agricultural or recreational use.

Most of the ferrous sulphate wastes are treated with lime to make them alkaline. The solutions are then transferred to lagoons where soluble ferrous hydroxide oxidises to ferric hydroxide, which precipitates, aided somewhat by water evaporation. In this way, iron is disposed of by land storage and clean effluents are produced.

At some locations, the effluents are reused. The copper sulphate leach solution is electrolysed to deposit copper at the cathodes and the resultant regenerated sulphuric acid liquid is recycled. In this way, production of iron sulphate waste is avoided.

Integrated Waste Treatment for Primary Production

Where more than one Wastewater is produced, the possibility of combining them in some way should be examined. An example of integrated treatment, using leaching wastes and ore tailings, is used by various companies.

The essential feature of this joint treatment is a combination of the alkaline tailing waste with the acidic leaching waste to form a neutral iron-free effluent for discharge. Two tailing waste sources are used: sand and slime sediment formed from lagooning tailing wastes in previous years and Wastewater from currently used tailing disposal lagoons; both contain residual lime.

The leaching Wastewater, having a pH of 4, first passes through the old tailings lagoon where it is partially neutralised by contact with the residual lime. By thus passing over the area, it controls dusting which otherwise presents an air pollution problem.

After collection, the water flows through a ditch where clear water from the active tailing lagoon, of pH 11, is added to produce a pH of about 7. At times of high leach flow or cold weather, milk of lime is added to complete neutralisation.

This combined stream flows into a 400 acre settling pond which permits oxidisation of the ferrous hydroxide to ferric hydroxide. In settling, ferric hydroxide carries down with it any other solids suspended in the stream.

Pickle Rinse Waters from Fabrication

At plants where pickle bath rinse waters and dumps are segregated from other wastes, rinse waters have been treated successfully by neutralisation and precipitation. A good example of segregated wastes treated in this way, where rinse waters contain 10-20 mg/l each of copper and zinc and have a pH of about 2.5.

To even out large fluctuations in flow rate that occur (0-800 gpm), these dilute wastes are first collected in an equalisation lagoon; based on the average flow rate of 400 gpm, this lagoon has an 8-hour capacity.

A steady 400 gpm flow-from the equalisation lagoon is pumped to a 1000 gal mixing tank where slaked lime is added to neutralise the acid and to raise the pH to 11-12. The spent pickle liquor, when dumped, is sent to a separate storage tank from which it is pumped slowly, over several days, to the mixing tank where it is combined with rinse waters.

These strong wastes are dumped when the mill is shut down, using the same pump and parts of the piping used for rinse waters. Provision is made for a pre-treatment reduction to the trivalent form of the hexavalent chromium in dichromate pickle liquors; this is done by addition of sodium bisulphite and acid to the liquors in the storage tank.

The high pH liquid from the mixing tank is separated in a clariflocculator, composed of a flocculator in the center of an annular clarifier; at average flowrates, these two parts have detention times of 20 and 170 minutes, respectively. The slurry underflow is discharged to one of two sludge lagoons, of 250 day combined capacity; top water from the sludge lagoon is returned to the equalisation lagoon and compacted sludge is removed periodically.

The clear effluent of pH 11-12 and containing 1-2 mg/l each of copper and zinc is diluted fourfold by mixing with other process water effluent and discharged to the nearby river. These treatment facilities have for several years produced a final effluent containing acceptable levels of copper, zinc and pH.

Attempts to reduce the size of treatment facilities by process changes in the mills themselves have been successful. In one approach, by changes in cycle time and drain angles, more of the pickle liquor is allowed to drain back into the acid bath before rinsing.

This improvement in rinse procedure results in a decrease in the amount of acid waste produced and a reduction in the flow rate of rinse water required, both of which allowed construction of a smaller waste treatment plant than would otherwise have been possible.

The other approach also involved rinse procedures; countercurrent rinses were used and the rinse water flow rate was controlled by pH. As a result, the wastes produced are fairly consistent in concentration, so smaller variations in lime addition are required; fluctuations in flow are decreased the equalising lagoon.

Metal Fabricating Plants

Generally the Wastewaters originating in metal fabricating plants are similar regardless of whether they originate from small shops or large production facilities. For the most part the Wastewaters from such plants contain small amounts of metal particles, free and soluble oils, and various cleaning compounds used in cleaning either the product or the shop itself.

Some plants have Wastewaters from air pollution control devices for painting operations. These residues are treated in a manner similar to free oils. Normally, the wastes are slightly acid or alkaline, and they are usually opaque, milk-coloured, and contain some free (non-soluble) oil.

The simplest form of treating such wastes is by means of gravity skimming tanks. In many cases such skimming will be sufficient to permit discharge of the industrial waste to a sewer. Optimum gravity skimming tanks are generally designed along standards set forth by the American Petroleum Institute which relate tank dimensions to particle size, oil rise rates and detention time as well as to the nature of the type of oil to be skimmed.

Such tanks operate best when supplied with a limited, uniform influent flow rate. As a result it is often wise to optimise the skimming tank design and then utilise large holding or equalising tanks upstream of the actual skimming tanks to permit pumps to deliver the Wastewater to the gravity skimmers at a fixed design rate.

The gravity skimming will remove free oils. The metal particles will settle in the bottom of the tank and can be removed manually at infrequent intervals or automatically by means of drag conveyors if such metal particles are deposited in significant quantities.

If the metal fabricating plant Wastewater contains considerable soluble oils, generally used as a machining coolant, further treatment will be required in addition to gravity skimming. Usually wastewaters from a metal fabricating plant will be contaminated with soluble oils, since even if soluble oils are not used in plant processes, they may well develop because of the mixing of free oils and metal cleaners or emulsifying agents when brought together in the Wastewater collection system.

Batch Treatment

Under these conditions a basic decision must be made when considering such waste treatment – whether or not the qualities and quantities of the Wastewaters are such that a batch system is in order of whether a constant flow system utilising continuous skimming and other oil removal methods would be preferred.

Generally speaking, small quantities of less than 2000 gpd can best be handled by batch type systems, with larger quantities being treated by a continuous system. However, in the event there is a possibility of toxic contaminants such as plating Wastewaters, a batch treatment system becomes essential.

A batch type system more readily permits testing and additional treatment to be applied to the wastewaters if this becomes necessary. Certain cleaning compounds can mix with the soluble oil waste and convert it into a jellylike material which will defy treatment by usual methods. Such complications can be accommodated only by a batch system.

These treatment systems consist of parallel batch collection and skimming tanks. When sufficient Wastewaters have been accumulated in one batch tank, the free oil is skimmed and conveyed to a waste oil collection tank.

Miscellaneous grit and metal particles can be allowed to accumulate in the batch tanks until manual cleaning is required. The remaining Wastewaters containing soluble oils are then cracked in either the batch tank or a separate retention tank to break the oil emulsion.

This is usually done by means of adding acid or alum until the pH has been lowered sufficiently to cause the emulsion to break down. The free oil is then skimmed or decanted in a separator and the remaining waters are neutralised by means of caustic soda or lime.

If sodium hydroxide (caustic soda) is utilised the clarified Wastewater may be high in dissolved solids consisting mainly of the soluble sodium salts resulting from neutralisation. Frequently such waste cannot be discharged directly to a stream, but it is usually acceptable in municipal sewer systems.

If lime is used for neutralisation the Wastewaters will contain few dissolved solids because of the relative insolubility of the resulting calcium salts. However, neutralisation with lime will precipitate a sludge consisting of the calcium salts of the acid. This presents the problem of disposal of the resulting sludge.

This sludge can be collected and stored in a holding tank for ultimate disposal. Not infrequently it is wise to dry this sludge on vacuum filters, since with its high water content it can present a difficult problem with respect to storage and/or disposal of the precipitate.

Removal of most of the water will reduce the volume of the precipitate or sludge to a more easily handled quantity. While lime neutralisation of acid Wastewaters is less expensive than caustic soda neutralisation, the cost of sludge handling or drying may far outweigh the savings achieved by the use of lime. The volume and concentrations of Wastewaters must be known and the resulting sludge volumes calculated before a comparative economic study can be made.

Several new proprietary compounds have recently been marketed which will crack oil emulsions. Such chemicals usually will not require subsequent neutralisation with possible resulting sludge complications.

Continuous Treatment

Perhaps the most successful of all metal fabricating plant continuous waste treatment systems are those involving air flotation. In this process the soluble oil contaminated Wastewaters are collected in large hold of equalising tanks.

This permits one-shift operation of the air flotation unit at fixed input rates. Operation for three shifts would permit smaller hold tanks to be used but might not provide the quantitative or qualitative equalising of the Wastewaters.

After collection the Wastewaters are conveyed by transfer pumps to the retention tank. Acid, alkali and soda ash are added as indicated, and the waste is detained in the retention tank under air pressure long enough to absorb quantities of air.

Upon release through a pressure regulator valve into the air flotation unit the cracked soluble oil is floated to the top surface along with the minute floe quantities and swept into a sludge hold tank by scrappers or flight conveyors. The relatively clear effluent can then be disposed of as indicated and the heavy oil filled floe disposed of by incineration or tank wagon disposal methods.

The effluent may require further filtration or sedimentation in detention ponds to meet stream requirements. The less stringent requirements for discharges to sanitary sewers will usually obviate needs for filters or detention ponds.

Incineration of resulting precipitates or sludge is an excellent method of disposing of this waste product. Incinerators are available which develop temperatures high enough to burn the residues completely and which will present few air pollution hazards. In the event toxic material such as the metal salts of plating baths is in the sludge, this method is most desirable.

The oil content of the sludge fed into such a device is generally high enough to be self-sustaining in combustion: however, it is always necessary to provide supplementary fuel such as gas or oil to permit the incinerator to be heated initially to a high enough temperature to destroy the waste product when it is first introduced into the incinerator. If the water content of the sludge is high or the oil content is low, continuous use of supplementary fuel may be required.

The other means of continuously removing contaminants from metal fabricating plant Wastewater is by means of automating a batch treatment system, e.g. settling out the metal particles, skimming the free oils, chemically cracking the soluble oils and again skimming, neutralising the now oil free waters using caustic soda or lime, and dumping the clarified waters into a sewer or stream.

This amounts to a continuous flowing, instrumented version of the batch system. This would require pH sensing devices in the batch tanks to regulate the acid pump. Level switches in the batch tanks would be needed to activate the pump to the retention tank and the alum pump.

Also, automatic monitoring and recording of the final effluent properties would be required. There are many points in this type of system which must be instrumented to achieve the required automatic control using pH and level sensing devices, and it may therefore be difficult to achieve reliability.

The clinging of residual oils and/or sludge to the electrodes of pH sensing devices and the possibilities of unexplained materials finding their way into the system and fouling the chemical treatment and sensing devices almost preclude a trouble free automatic system.

As a result, for most plants the most practical approach to an automatic system is the air flotation system. Several Wastewater treatment equipment manufacturers construct such devices, and in general most of them are satisfactory.

However, care must be taken not to overload an air flotation machine. A slight overload of the machine will cause substantial reductions in the pollutant removal efficiencies. A 10 per cent hydraulic overload may cause as much as 50 per cent reduction in oil removal efficiencies and result in carryover of oil bearing floe with the clarified effluent.

All automatic or batch systems must be examined to be sure that valves, piping materials, tanks, coatings and linings are suitable for the oils, pH variations and chemicals used in the waste treatment process. Such considerations will be important to the cost of the facilities.

System Sizing

As in all industrial waste treatment problems, the question of anticipating waste quantities and strengths in conjunction with the design of a new metal fabricating Wastewater treatment facility is much more difficult than measuring the known quantities of properties of a waste that occurs in existing facilities.

As a result, the Wastewater treatment facility for a new proposed metal fabricating plant must be designed much more conservatively than one for an existing plant in which the effluent quantities and properties can be measured.

A possible solution, if it can be achieved, is to go into production on a new facility and, once waste quantities are determined, design the Wastewater treatment equipment. This takes cooperation on the part of state and regulatory agencies and may require tank wagon disposal or scavenger service for the first few months of operation.

Since Wastewater treatment is usually an overhead cost, all possible means to defray expenses must be examined. In the case of metal fabricating plant Wastewaters, several possibilities exist. The recovered waste oil may be reused or at least sold for road oil, and the treated Wastewater may be reused to satisfy plant process, cooling, or other non-potable water demands.

Carried to the ultimate end, such reuse may virtually eliminate plant Wastewater discharges, thereby negating the involvement with Wastewater regulatory agencies.

Thus, to make greater reuse of water possible, more stringent requirements on quality of the effluent to the nation's waterways will be enforced on cities and industries. Some relief will be accomplished by better water planning, for instance, by storing of water from periods of high precipitation.

Research will be required to develop improved methods of treating sanitary wastes. Economical means are required to remove a higher percentage of pollutants. However, if the techniques now available were applied, a vast improvement in the nation's waterways would result.

Wastewater Treatment Challenges in Food Processing

Wastewater generated from food production and agricultural activities is a major source of environmental pollution. It is also among the most difficult and costly waste to manage because food processing wastewater can contain large quantities of nutrients, organic carbon, nitrogenous organics, inorganics, suspended and dissolved solids, and it has high biochemical and chemical oxygen demands. It must be treated to levels that will not damage receiving waters due to excessive nutrients or oxygen demand when directly discharged or will not disrupt publicly owned treatment works (POTWs) when discharged to sewers. Plant-food processing wastes may be lower strength and greater volume than animal processing and animal production.

"Each type of food processing wastewater will have special factors to consider, and in addition to the technology performance issues, seasonality of production adds to the complexities of the treatment choices and operations in several industries".

The range of food and agricultural wastes present different challenges. Industry examples include: meat and poultry products, dairy products, fruits and vegetables for canning and preserving, grain products, sugar and related confectionaries, fats and oils, and beverages and brewing, among others. Biochemical oxygen demand (BOD) and chemical oxygen demand (COD) values for many wastes are in the thousands of milligrams per liter, and some like cheese production, winery and olive milling can be in the tens of thousands for COD. So, each waste type will have special factors to consider, and in addition to the technology performance issues, seasonality of production adds to the complexities of the treatment choices and operations in several industries.

Treatment Technologies for Food Processing Wastewater

The types of treatment technologies used for food processing wastewater are not unusual among wastewater treatment options and include the typical array of biological and physical chemical treatments. Both oxidative and anaerobic processes may be employed. They include: flotation, coagulation, sedimentation, filtration, adsorption, membranes, primary settling, secondary activated sludge, anaerobic digestion and even recovery of carbon dioxide or methane for subsequent uses. In addition, many wastes or treated residues are amenable for land application as beneficial soil amendments or fertilizers, which can mitigate much of the wastewater discharge concerns; however, even those approaches can be problematic if excessive nutrient runoff occurs, particularly in cold weather.

"The safety of the food product is paramount, and recycling within a process step can be allowed by law for certain food processes, such as chiller water in meat and poultry processing."

In-plant recycling can also be employed since water is in demand for not only processing steps, but also for cleaning equipment, facilities and floors. Detergents or other additives may be employed for those nonfood contact applications, which may improve or complicate the subsequent treatment processes. The safety of the food product is paramount, and recycling within a process step can be allowed by law for certain food processes, such as chiller water in meat and poultry processing. Recycling where food contact is involved is more problematic between processes because different contaminants may be contributed.

Fruit and Vegetable Processing

Many of the aforementioned treatment processes are used in vegetable and food processing wastewater. The production process produces waste streams from washing and rinsing, sorting, in-plant fluid transport methods, peeling, pureeing and juicing, blanching, canning, drying, cooking and cleanup. Most of the waste content is biodegradable carbohydrates, although salts may also be a contributor for some such as brining products.

Fishing Industry

There are many elements of production of fish products that generate solid and liquid wastes. Harvested product may be processed on shipboard or stored by icing or freezing for transport to the processing plant. Farmed product may be handled somewhat differently from wild-caught product. Wastes from eviscerating and butchering are collected, dried or screened and used as by-products. The processing plant may further trim from the product and then cook, pack or freeze, generating other waste liquids and solids. Each type of product generates different levels of BOD, COD and often substantial amounts of fats, oils and grease and protein content.

Meat and Poultry Industries

Wastes from the meat and poultry industries include those generated by the animals during live-stock holding as well as high-strength wastes produced during processing. Processes include: slaughtering, defeathering or hide removal, eviscerating and trimming, washing, disinfecting and cooling. Some poultry plants may process 2 million birds per week, so they are large-scale waste generators. Nitrogenous organic (proteins), fats and inorganic (nitrates) waste components are substantial.

Dairy Production

The two main elements are milk bottling and milk product production of whey, butter, cheese, ice cream, yogurts, cottage cheese and other milk derivatives. The latter post-milk processing is the largest contributor to wastewater production and to the strength of the wastewater. The COD levels at those stages, especially in cheese production, can be much more than 10 times the amount in milk bottling. Nondairy ingredients such as flavors, sugars and fruits are also involved in production, and they can contribute to the waste stream. The waste products are mainly biodegradable, so aerobic and anaerobic processes are standard.

Freila.

Oil and Fat Processing

Vegetable oil and animal fat production often involves solvent extraction or compression for olive and sesame oils. The oils must have further refining processes to remove some taste, free fatty acids and other residue components from the extracted oils and fats. Substantial amounts of sol-id residues are also generated from the vegetable oil production. In addition, extraction solvents must be removed. Olive oil production generates higher levels of COD and solid wastes compared to other food and agricultural wastes.

Anaerobic Treatment for Food Processing Wastewater

Anaerobic technologies are a growing area of interest in agricultural and food processing wastewater treatment because of the opportunity to generate methane gas as a byproduct, which can be used to generate heat and electrical energy to offset facility operating costs, and it reduces the biological activity and the volume of waste and the carbon footprint. Anaerobic processes are often relatively slow and temperature sensitive, and they require larger facilities, so the goal is to develop processes that are more rapid to optimize throughput and methane production. Anaerobic methane and energy production are also being more commonly used in municipal waste treatment plants to provide heat and electricity. The upfront investments are substantial, and they require more sophisticated operations management, but they are providing POTWs with longer-term insulation from increasing energy costs. Food wastes are even used to supplement POTW waste to increase methane and energy production.

"Food and agricultural wastes are ideal for biogas production due to the relatively high total organic carbon loadings compared to many other wastes."

Food and agricultural wastes are ideal for biogas production due to the relatively high total organic carbon loadings compared to many other wastes. For example, a large municipal plant in Washington, D.C., obtains about one-third of its electricity needs using its Cambi thermal hydrolysis, advanced anaerobic digestion and biogas process. Also, an Oakland, California, facility, which is likely a recipient of substantial food-processing wastes, is reported to produce excess electricity that it can sell to the grid.

High-rate anaerobic digesters are attracting interest because of their higher loading capacities and lower sludge production. They can include: anaerobic filters, upflow anaerobic sludge blanket reactors, baffled fluidized beds, granular sludge beds, sequencing batch reactors and hybrid/hybrid upflow sludge blanket reactors. An important consideration is the presence of granular support media that provide an enlarged surface area, which provides enhanced contact between the active microbial species on the surface and carbonaceous material and nutrients in the wastewater. Refractory waste materials can become more biologically available and more productive by use of appropriate pretreatments. There is also exploration of inoculating with anammox bacteria to accelerate conversion of nitrogenous materials. Some POTWs are using these to improve the efficiency and reduce costs of traditional nitrification/denitrification processes.

Discharges of food and agricultural wastes are a significant contributor to nutrient and carbonaceous and nitrogenous waste discharges. Treatment of agricultural and food processing wastewater is complex and costly because of the contaminant loadings and the variability of the different wastes encountered in a plant. Industries including poultry and meat processing, dairy products and oil production generate high-strength wastes. While common wastewater treatment processes are used, there are developments in anaerobic processes to produce methane gas for energy and electricity to offset process costs. In addition to reducing operating costs, they are environmentally friendly by reducing waste discharges and carbon footprints.

Dairy Wastewater Treatment

The dairy industry includes the transformation of raw milk into pasteurised and sour milk, yoghurt, hard, soft and cottage cheese, cream and butter products, ice cream, milk and whey powders, lactose, condensed milk, as well as various types of desserts. The general distinctions among these foods are due to the reuse of non- -fat milk and whey (a by-product in cheese manufacturing) and the evaporation of the free water from the coagulum as well as from milk and whey powders. With the rapid industrialisation observed in the last century and the growing rate of milk production (around 2.8% per annum), dairy processing is usually considered the largest industrial food wastewater source, especially in Europe. Moreover, in around 50% of the world's whey production, especially concerning acid whey, it is untreated prior to disposal. The effluents originating from various production technologies are not discharged simultaneously, thus forming a stream with wide qualitative and quantitative variations. Notwithstanding the differences in composition, attributable to the manufactured product and technological operations, dairy effluents are distinguished by their relatively increased temperature, high organic content and a wide pH range, which requires special purification in order to eliminate or reduce environmental damage. Treatments of dairy wastewaters include the application of mechanical, physicochemical and biological methods. Mechanical treatment is necessary to equalise volumetric and mass flow changes. It also reduces parts of the suspended solids. Physicochemical processes are effective in the removal of emulsified compounds, but reagent addition increases water treatment costs. Another disadvantage is the very low elimination of soluble chemical oxygen demand (COD). Therefore, biological wastewater treatment systems are preferred due to the highly biodegradable contaminants.

Dairy Wastewater Characteristics

Wastewater Volume

Water plays a key role in milk processing. It is used in every step of the technological lines, including cleaning and washing, disinfection, heating and cooling. Water requirements are huge.

The bulk of wastewater comes from manufacturing processes. Contaminated water, including sanitary activities, reaches 50–80% of the total water consumed in the dairy factory, whereas the remaining 20–50% is conditionally clean. It has been estimated that the amount of wastewater is approx. 2.5 times higher than that of processed milk in units of volume. The amount and characteristics of the wastewater depend largely on the factory size, applied technology, effectiveness and complexity of clean-in-place (CIP) methods, good manufacture practices (GMP), *etc.* However, the introduction of GMP can reduce the world's wastewater mean volume from 0.5–37 to 0.5–2 m^3 of effluent per m^3 of processed milk. Nowadays, the designed volumetric load is 1 m^3 of effluent per tonne of manufactured milk.

In dairy plants, the great fluctuations in wastewater quality and quantity are very problematic because each milk product needs a separate technological line. This results in the change of dairy effluent composition with the start of a new cycle in the manufacturing process, which impedes the work of in-factory wastewater treatment plants. Furthermore, intensive effluent volumetric variations in time are commonly observed. Daily and hourly changes are the consequence of washing the equipment and floors as the final step in every process cycle. Seasonal variations can be attributed to a higher dairy plant load in summer than in winter. One way of explaining hourly homogeneity is by coefficients in the range of 1.4–2.0. The diurnal inequality coefficient depends on the seasonal character of dairy processing, varying from 1.5 for 2- and 3-shift work in summer to 2.6 for winter shifts. The actual concentration of polluting dairy effluents varies widely depending on the profile and capacity of the company, the production technology, the type of equipment used, the degree of wastewater reuse, the loss of raw materials, waste management, *etc.* A major factor in the volumetric loading of dairy wastewater treatment plants are the immediate discharges produced in the cleaning of tank trucks, pipelines or equipment at the end of each cycle. In such cases, the effluent volumes are higher than those of manufactured milk. On average, wastewater discharge is 70% of the amount of the fresh water used at the plant.

Dairy processing effluents mostly include milk or milk products lost in the technological cycles (spilled milk, spoiled milk, skimmed milk and curd pieces); starter cultures used in manufacturing; by-products of processing operations (whey, milk and whey permeates); contaminants from the washing of milk trucks, tanks, cans, equipment, bottles and floors; reagents applied in CIP procedures, cooling of milk and milk products, for sanitary needs, in equipment damage or operational problems; and various additives introduced in manufacturing. Milk loss in wastewater is around 0.5–2.5% of processed milk, but it can increase to 3–4%.

Wastewater Categories

The wide variety of dairy products presupposes the existence of many wastewater types. However, three major categories can be outlined according to their origin and composition:

Processing Water

Processing water is formed in the cooling of milk in special coolers and condensers, as well as condensates from the evaporation of milk or whey. Milk and whey drying produces vapours which form the cleanest effluent after condensation although it may contain volatile substances as well as milk or whey droplets from evaporators. In general, processing waters lack pollutants and, after minimal pretreatment, they can be reused or discharged together with stormwater. Water reusage is possible for installations that are not in direct contact with derived products. Typical applications include hot water and steam production as well as membrane cleaning. The water from the cooling of products during pasteurisation after the last rinse of bottles and condensates generated in vacuum installations from secondary vapours can be utilised for room cleaning, lawn irrigation, etc.

Cleaning Wastewater

Cleaning wastewater usually comes from washing equipment which is in direct contact with milk or dairy products. It also includes milk and product spillage, whey, pressing and brine, CIP effluents or equipment malfunction and even operational errors. Over 90% of organic solids in effluents come from milk and manufacturing residues: cheese pieces, whey, cream, water from separation and clarification, starter cultures, yoghurt, fruit concentrates or stabilisers. These effluents are in large quantities and are highly polluted, thus requiring further treatment.

Sanitary Wastewater

Sanitary wastewater is found in lavatories, shower rooms, *etc.* Sanitary wastewater is similar in composition to municipal wastewater and is generally piped directly to sewage works. It can be used as nitrogen source for unbalanced dairy effluents before a secondary aerobic treatment.

Additionally, the by-products of manufacturing processes, such as whey, milk and whey permeates, can also be grouped separately if they are collected individually from other wastewater streams.

The main pollutant in milk processing wastewater is whey due to its high organic and volumetric load. It represents about 85–95% of the milk volume and 55% of the milk components. Whey consists of carbohydrates (4–5%), mostly lactose. Proteins and lactic acid amount to less than 1%, fats to around 0.4–0.5%, while salts vary from 1 to 3%. Whey is produced mainly in cheese manufacturing, and its volume depends on the productivity of cheese and the type of processed milk – bovine, goat, sheep, etc. On the basis of milk casein coagulation procedures, whey can be categorised as cheese whey and second cheese whey. Cheese whey is a by-product in the production of hard, semi-hard and soft cheese, after the addition of rennin to milk. Mild enzyme action produces sweet whey with a pH=6–7. Second cheese whey is a by-product in cottage cheese production after milk has been fermented, or curdled, with organic or mineral acids. Due to strong acid conditions, whey develops an acidic taste, while the average pH value rarely exceeds 5. Scientific literature also discusses casein whey whose composition is very close to that of second cheese whey. Sweet and acid whey also differ in mineral and protein content.

During cheese manufacturing, cheese whey wastewater is produced as well. Its volume and composition change with respect to the type of produced cheese, the applied technology, the milk type and the environment. Cheese whey wastewater originates in the addition of surplus cheese whey and second cheese whey to washing effluents. Nevertheless, its contamination level is lower than that of cheese whey.

Cheese whey waste streams are valuable sources of different compounds (protein, lactose, mineral elements) and are utilised in the manufacture of various products, such as lactic acid, single-cell protein, baker's yeast, starter cultures, fermented whey drinks, enzymes, antibiotics, organic acids, vitamins, food gums, etc. Nevertheless, it should be taken into account that whey or whey product recovery results in new waste streams which also need to be treated although such effluents are less polluted than whey and their organic loading is comparable to other dairy wastewater.

Milk and whey permeates are by-products in cheese manufacturing; they are produced during milk and whey ultrafiltration, respectively. Their solid content is lower; they are rich in soluble compounds, over 80% of which is lactose.

Dairy wastewater consists of complex constituents. Knowing the composition of milk and milk products, we can estimate better the wastewater contaminant loading. Although milk manufacturing produces waste streams analogous to milk and dairy product loss, every process gives an effluent unique in volume and composition.

Dairy wastewater volumetric and flow rates (depending on the production capacity and work shifts), as well as pH and total suspended solids (TSS) content (as a consequence of applied CIP methods) affect the efficiency of wastewater treatment management. It is important to know the quantity of the milk to be pasteurised, how much milk is processed into cheese and whether the entire obtained whey is discharged in wastewater or part of it is processed and reused. Contaminant concentrations in wastewater can be determined by using:

$$C\text{-}(L_2 \cdot C_1 + L_2 \cdot C_2 + \ldots + L_n \cdot C_n)/N_1 + N_2 + \ldots + N_n)$$

where C is contaminant concentration in wastewater (g/m³), L is the loss of milk and milk products in different technological production cycles expressed in proportional units (m³ or t), C_1, C_2 and C_n are contaminant concentration per unit of milk or milk product loss (g/t), and N_1, N_2 and N_n are wastewater discharge per unit of milk or generated milk products (m³/t).

Dairy Wastewater Composition

Milk processing effluents have an increased temperature and large variations in pH, TSS, biological oxygen demand (BOD), COD, total nitrogen (TN), total phosphorus (TP) and fat, oil and grease (FOG). There is little information on industrial-scale dairy effluent composition.

Typically, dairy wastewater is white in colour (whey is yellowish-green) and has an unpleasant odour and turbid character.

With annual temperatures of 17–25 °C, dairy waste streams are warmer than municipal wastewater (10–20 °C), which results in faster biological degradation compared to sewage treatment plants. The average temperatures of industrial dairy effluents range from 17–18 °C in winter and 22–25 °C in summer. Using the Arrhenius equation, the biodegradation rates and oxygen consumption can be predicted to be 1.5 times higher in summer than in winter. The design winter temperature of 15 °C is adopted for this type of wastewater due to the utilisation of hot water for washing and cleaning of equipment.

A crucial requirement for biological treatment of dairy wastewater is their pH value between 6 and 9. Milk and butter factories have effluents with active reaction close to neutral (pH=6.8–7.4). In plants where a certain amount of whey is discharged, the pH of the effluent is reduced to below 6.2. In cheese manufacturing, sweet whey is slightly acidic, with pH=5.9–6.6, while mineral acid coagulation gives an acidic whey with pH=4.3–4.6. The sharp increase in the short-term pH of the total flow of up to 10–10.5 is attributable to the discharge of alkaline cleaning solutions. The prolonged exposure of wastewater to anaerobic conditions (in the sewer network with sumps) causes liquid acidification by lactic acid fermentation that leads to a decrease in pH.

Although dairy wastewaters have low concentrations of settleable solids, they may clog sewage pipes. Most of the suspension enters the initial stage of equipment cleaning. The bulk of the sediment (90%) of organic matter is usually of protein origin, namely particles of solid milk processing (pieces of cheese, coagulated milk, cheese, curd fines, milk film or flavouring agents, *etc.*) and other impurities (soil or sand) that get into the sewage system during equipment washing or packaging. Formation of protein and fat deposits on the inside of the pipes requires periodic cleaning with appropriate chemical or bacterial preparations. The main advantage in the application of such bacteria is that they continue acting in the next stages of wastewater treatment, increasing the purification effect. The highest amount of total solids (TS) has been reported in whey, with negligible amount of volatiles. Fats in dairy industry effluents are found in trace amounts in the form of emulsions with a droplet diameter of 1–10 μm. During homogenisation, the size of milk fat globules is reduced to 1–2 μm. The obtained stable emulsion, when passing into dairy effluents, affects the mechanical wastewater treatment system due to its difficult separation. Thus, fats remaining in cheese whey wastewater can produce an undesired flotation, which results in the washout of active sludge during biological processes. In the production of high-fat products (cream, sour cream and butter), larger fat globules are extracted from the milk, due to their coalescence and enlargement, as well as the degradation of the protein shell. That is why fat impurities in the wastewater from these productions are significantly different in type and concentration and their elimination by settling is more efficient than in other dairy effluents. The FOG concentration in the wastewater from dairy plants specialised in the production of high-fat products is 0.2–0.4 g/L although higher values (up to 2.88 g/L in a butter factory have been reported). In the wastewater from other dairy plants, it usually does not exceed 0.1 g/L.

Due to their high organic content, represented mainly by rapidly assimilable carbohydrates and slowly degradable proteins and lipids, dairy wastewater is characterised by high BOD and COD values varying from 0.1 to 100 g/L. It is known that there is a direct relationship between the ultimate 20-day BOD (BOD_u) and COD values in dairy wastewater, as shown in:

$$BOD_0 = (0.80 - 0.84) \cdot COD.$$

It should be taken into account that such a logical connection cannot be made between a 5-day BOD (BOD_5) and BOD_u, and between BOD_5 and COD. Therefore, BOD_5 value of dairy waste streams is not an objective indicator of organic pollution. Nevertheless, many authors use the BOD_5 value of dairy wastewater in the BOD_5/COD ratio. For dairy effluents this ratio varies between 0.4 and 0.8. However, it should be determined separately in every particular case and, since dairy wastewater is industrial, the BOD analysis should be conducted with selected microbial consortia, instead of traditional seeding material in order to achieve reliable results.

The highest whey COD and BOD_5 concentrations have been reported to be between 60–80 and 30–50 g/L, respectively. About 90% of BOD and COD loading is caused by lactose, while protein removal contributes to only around 12% of the whey COD reduction. High lactose solubility increases soluble COD part, which is removed mostly by biological units. Like whey, milk and whey permeates have high COD load because they are rich in lactose, which excludes the possibility for a direct discharge in water bodies. Cheese whey wastewater also has increased concentrations of organic matter, the values varying significantly: 0.8–77 g/L of COD and 0.6–16 g/L of BOD_5. The lower lactose concentration reported is due to the fermentation in anaerobic conditions that leads to a lower initial pH and casein precipitation and odour production from the obtained butyric acid.

The time-consuming BOD analysis requires the application of faster methods that determine aerobically digestible organic matter in dairy wastewater. Many authors show that COD fractionation combined with the calculation of respirometric oxygen uptake rate is a good alternative method for the determination of wastewater biodegradability. However, results were obtained only for mixed dairy wastewater, while the information on single manufacturing processes is insufficient. Total organic carbon (TOC) calculation also includes organic carbonaceous fractions. It gives immediate results and can be used for online measurements. However, the TOC-BOD relationship should be estimated first. There is no available scientific data for the online TOC application or the TOC-BOD relationship in dairy wastewater treatment.

Every milk effluent has notably different TN and TP concentrations. Nitrogen exists mainly in the form of amino groups from milk proteins. Other nitrogenous compounds are also detected: urea, uric acids, and NH_4^+, NO_2^- and NO_3^- ions. Small quantities of nitrogen ammonium salts originating from ammonia compressors can also be found. Phosphorus compounds are mostly inorganic, phosphate (PO_4^{3-}) and diphosphate ($P_2O_7^{4-}$), but they can also be present in organic form. Total nitrogen content in the wastewater from urban dairies, dairy and butter plants is 4.2–6% and that from cheese factories 3.7% of the BOD_5. The phosphorus concentration is in the 0.6–0.7% range of the BOD_5. The reported TN and TP values demonstrate an increased eutrophication risk in water receivers. Their concentrations are sufficient for normal biological treatment processes and the respective growth of bacteria involved in the oxidation of dairy wastewater impurities. However, cheese effluents lack in nitrogen for proper aerobic biological treatment due to the following C/N/P ratio of approx. 200:3.5:1 but can easily be treated anaerobically. During biological treatment of cheese factory wastewater, nitrification is less intense than in other dairy industry wastewater treatment facilities because of the lower BOD_5/N ratio.

Dairy effluents are characterised by very low alkalinity (approx. 2.5 g/L expressed as $CaCO_3$ in milk permeate), thus bringing about a potential for rapid acidification and increased reagent costs for pH maintenance during purification.

The high salinity of industrial dairy effluents causes a non-volatile suspended solid content increase in the primary and secondary sludge. Inorganic impurities in dairy wastewater are represented by Na^+, K^+, Ca^{2+} and Cl^- ions, with their highest amounts in cheese and cottage cheese production (0.46–10%), mostly NaCl and KCl (>50%) as well as $Ca_3(PO_4)_2$, where salt is added in advance. Increased Na^+ amounts indicate the application of alkaline cleaning agents in milk factories. The amount of Ca^{2+} in acidic whey is twice as high as that in sweet whey. The presence of chlorides in dairy wastewater is due to the addition of salt in the production of brine and cooling liquors, and the Cl^- concentration in fresh water and milk. Cl^- concentration in dairy wastewater reaches 0.8–1 g/L but the average value range is 0.15–0.2 g/L.

The additional wastewater pollution due to the used cleaning solutions, additives and other products which enter the drainage pipes should be taken into account. CIP methods produce wastewater streams at 12- or 24-hour time intervals, while sanitisers are used if the dairy factory has been shut down for more than 96 h. Thus, wastewater pH will change widely depending on the cleaning program applied. Different chemical solutions can be used in accordance with the installation type, water hardness, etc. The cleaning agents applied in CIP procedures affect principally the effluent pH (mineral and organic acids), contributing less than 10% to BOD_5 and COD loading and increasing amounts of water for cleaning and disinfection (up to 30% of total water flow rate). Most of the applied chemicals are very toxic to microorganisms in secondary treatment units. NaOH and HCl increase the mineral scale, while HNO_3, quaternary ammonium surfactants, and detergents containing H_3PO_4 and P influence TN and TP loading, which leads to an accelerated eutrophication of the environment if not treated properly. Due to the above-mentioned environmental problems, the trend is to apply more HNO_3 instead of the less desirable H_3PO_4 although the latter is a better cleaner the application of which will not be reduced in the future. The cleaning solutions utilised in CIP procedures are hot (64–82 °C), which causes a temperature increase in the resulting effluents. Strong oxidants or bleaches (NaOCl and ClO_2) are applied for sanitising installations. Cl-containing bleaching agents can produce dangerous organochlorides which pollute dairy effluents. Enzymes as well as surfactants are the chemicals preferred for cool surface cleaning and cause fewer negative environmental problems. In minor doses, the following substances can also be found: NH_3, Na_3PO_4, HCl, $HOCH_2COOH$, Na_2SiO_3, hydraulic oil, propylene glycol, emulsifiers, antifoaming agents, sodium azide and chloramphenicol.

Dairy Wastewater Treatment

Dairy manufacturing has a strong impact on the environment, producing large volumes of wastewater with high organic and nutrient loading and extreme pH variations. This requires the application of effective and cheap wastewater treatment procedures which ensure fresh water preservation. There are various dairy effluent treatment strategies, which are described in the following paragraphs.

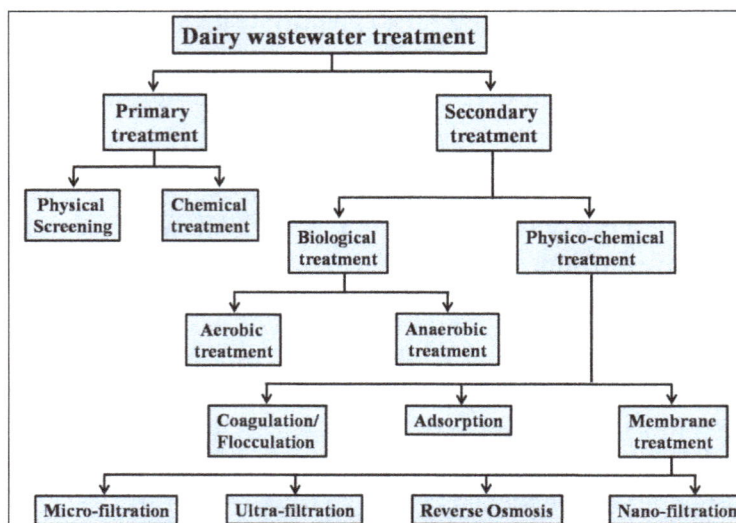

Dairy wastewater treatment options.

Discharge in Nature without Treatment

It is not recommended that raw dairy wastewater be discharged directly into water bodies because this would lead to different pollution problems, including rapid dissolved O_2 depletion due to the high organic loading, which results in anaerobic conditions, the release of volatile toxic substances, aquatic life destruction and subsequent environmental damage. Higher water temperatures decrease O_2 solubility and increase biota sensitivity.

Treatment in Wetlands

Wetland systems use natural processes that include self-supported microbial communities to improve wastewater treatment. The simple construction and the lack of sludge recycling make them preferable for dairy effluent utilisation in developing communities. The main drawbacks of their application include the need for a large surface area, the potential risks for surface and groundwater pollution, the presence of dangerous volatile substances and the presence of insects. The easy exploitation of the systems counteracts with the complexity of the biological processes, which exceeds that of other treatment systems applied in wastewater purification. Also problematic is the generation of Fe^{3+}, Mn^{3+} and Ca^{2+} ions. They precipitate and reduce bed permeability with time. As a result, anaerobic conditions prevail and the NH_3 removal is limited.

Generally, dairy wastewater is treated in wetlands under aerobic conditions. Five days are enough for an 85% BOD_5 reduction in aerobic ponds with milk wastes at 20 °C, while high-load dairy wastewater is treated mostly in facultative wetlands. In a butyl-covered lagoon, processing effluent was biodegraded at 35 °C, with the organic loading rate (OLR), expressed as COD, of 1.5 kg/(m³·-

day), neutral pH and a hydraulic retention time (HRT) of 1–2 days. However, a polishing step in aerated pond is necessary to achieve 99% total COD reduction. A surface-flow wetland was applied to utilise 2.65 m³/day of milkhouse wastewater with OLR, expressed as BOD_5, of 7.3 g/(m²·day). The results showed high TSS, BOD_5, TP and total Kjeldahl nitrogen (TKN) biodegradation with respective values of 94, 85, 68 and 53%. Despite the fact that the lagoon produced NH_3, its outflow concentrations gradually declined over time. Most of the nitrogen was stored in biomass, while denitrification had a minor role (<1%). Clarified effluents needed more BOD_5 reduction to meet water discharge standards. Cheese wastewater with OLR, expressed as COD, of 5.5 kg/(m³·day) was consistently treated in a grease trap, an upflow anaerobic sludge blanket (UASB) reactor-type pond, aerobic pond and a final wetland with water hyacinth. A high quality effluent was obtained: BOD_5, COD, TSS, FOG, organic N and total coliforms were reduced by more than 90%, except for the phosphorus from PO_4^{3-} (with a decrease of only 62%).

Purification in Urban or in-factory Wastewater Treatment Plant

In-plant effluent treatment is the most common strategy for dairy wastewater purification. Typically, it includes mechanical, physicochemical, chemical and biological methods.

Mechanical Treatment

Mechanical treatment removes suspended solids from wastewater. Conventional mechanical procedures reduce insufficiently the organic load because of the low settleable solid concentration in dairy wastewater. Nevertheless, the faster the wastewater is screened, the better, due to less TSS biodegradation and a low soluble COD increase.

High variations of dairy effluents can bring about an instability of the subsequent treatment facilities. Adequate equalisation will smooth the fluctuations in the flow, organic loading, pH and temperature, neutralise residual cleaning agents and completely destroy excess oxidisers. In practice, a 24-hour flow pattern at the highest load can be effectively handled by effluent equalisation for at least 6–12 h with a basin dimension from 25 to 50% of the total effluent volume.

Physicochemical Treatment

Physicochemical treatment destroys and reduces milk fat and protein colloids in the dairy wastewater. FOG removal is a major problem in the plants producing unskimmed milk, in milk and whey separation, cheese and butter production, as well as milk bottling. Skimmed milk production rarely creates such problems.

Animal fat is solid at room temperature due to the high levels of saturated fatty acids in its composition. Milk fat is no exception. This physical state, combined with the low density of the fat allows its easy removal from the surface of wastewater. If the equalisation unit precedes the FOG trap, a temperature drop will heighten the risk of high fat accumulation on the top of the liquid. Otherwise, the equalisation unit must have a sufficient volume to collect the peak effluent flow. In general, flow balancing is followed by FOG removal. Increased wastewater temperatures can reduce fat separation ability. Dissolved air flotation is more effective because it reduces organic loading *via* protein and fat colloid destabilisation with coagulants ($Al_2(SO_4)_3$, $FeCl_3$ and $FeSO_4$) and flocculants. Nevertheless, this method requires expensive, synthetic chemicals which causes environmental problems and removes soluble matter to a lesser extent. The resulting scum is very hard to dewater and it is not recommended to mix it with activated sludge. Scum must be treated properly before disposal. If inorganic and synthetic chemicals are replaced by biopolymers (carboxymethyl cellulose (CMC) or chitosan), processed sludge can be used as an animal food ingredient.

According to some authors natural coagulation in dairy wastewater can be achieved with the application of certain lactic acid bacteria. These bacteria ferment soluble lactose to lactic acid, which denatures milk proteins in the wastewater. In combination with CMC, the total COD was reduced by 65–78%, while reduction of 49–82% was obtained when chitosan was used. At an initial 5 g/L of COD, over 0.01 g/L of proteins and 0.7–0.8 g/L of sugars, 75, >90 and 10–25% of COD were removed, respectively.

Chemical Treatment

Chemical treatment removes mostly colloids and soluble contaminants from milk processing effluents. It includes reagent oxidation or pH correction. During cheese wastewater reaction with $FeSO_4$ and H_2O_2, up to 80% of fat (initial concentration of 1.93 g/L) is removed. Extreme pH values of dairy wastewater below 6.5 and above 10 can increase the corrosion of pipes and be highly detrimental to microbiological assemblages in biological processes. Therefore, they should be corrected to reduce side effects. If a dissolved air flotation (DAF) unit is used, then the pH control is a necessary step to achieve optimal coagulant conditions. However, coagulants work best at an acidic pH, which requires a second pH adjustment to a neutral value before biological treatment. It is very suitable to collect independently used CIP solutions and outflow them constantly during the whole wastewater plant exploitation.

Biological Treatment

One of the most reliable methods for dairy effluent purification is biological removal. Such methods can assimilate all dairy wastewater components but they mostly utilise soluble compounds and small colloids. These processes have not been fully studied. Moreover, because of their unlimited adaptation potential, they can be jointly used in various sequences to meet certain component biodegradation requirements. Biological treatment has two main branches depending on oxygen requirements: aerobic and anaerobic processes.

Aerobic processes: Nowadays, most dairy wastewater treatment plants are aerobic although they have been less efficient, mainly due to filamentous growth and rapid acidification caused by high lactose levels and low water buffer capacity, respectively. Problems generally encountered with activated sludge processes are bulking and foaming, which diminish sludge settling, Fe^{3+} and CO_3^{2-} precipitation, additional biomass production as well as poor activity at low temperatures. It takes a few months for the sludge adaptation before full operational capacity is reached. Nitrogen from NH_3 is easily degraded. Phosphorus removal is less effective and relies on environmental conditions. Aerobic bacteria are less useful in colloid utilisation when compared to anaerobic bacteria. The heightened O_2 depletion (>3 kg of O_2 per kg of BOD_5) requires large energy demands during the aerobic treatment of concentrated dairy wastewater (>2 g of COD per L). Plug flow systems are better than complete-mix processes since they are less sensitive to high organic load problems like bulking sludge, *etc.* Commonly, dairy effluent OLR, expressed as BOD_5, should be less than

$0.28-0.30$ kg/m³. To enhance biological removal, a proper pretreatment or adequate wastewater dilution should be applied.

Aerobic biological systems give a very positive response during synthetic dairy wastewater treatment with 4 g/L of COD and 1 g/L of TKN at pH=11.5, with over 96% of degradation being achieved in a continuous mode. An artificial effluent similar to milk powder and butter processing wastewater was treated in an anaerobic- -anoxic-oxic system at HRT of 7 days and a nominal sludge age of 20 days. The process was characterised by sludge bulking due to the growth of filamentous bacteria (*Sphaerotilus natans*, Type 0411 and *Haliscomenobacter hydrossis*). TN removal remained unchanged at 66% without the improvement in the sludge volume index. TP depended on the anoxic selector relative dimensions (from 49 to 20%) and a respective nitrate rise in the effluent. Nevertheless, more than 90% of COD reduction was achieved.

Aerobic filters are applied to a lesser extent in the treatment of high-strength dairy effluents rich in FOG. High fat and heavy biofilm blockage are possible, which results in biomass loss, filter fouling and corresponding reduction in productivity.

The sequencing batch reactor (SBR) is preferred in dairy wastewater treatment because of its various loading capabilities and effluent flexibility. A traditional technology with free sludge flocs is mostly applied. The purification of milk effluents is given by Britz *et al*. COD was reduced by $91-97$, TS by 63, volatile solids (VS) by 66, TKN by 75, and TN by 38%. However, mechanical treatment had to be applied first. Another study shows the aerobic SBR as an excellent example of the combination of activated sludge granulation with dairy effluent treatment. Granulation stability is limited by nutrient concentration in the wastewater, while effluent quality depends on the need for preliminary sludge settling, usually $0.25-0.5$ HRT. Up to 90% of total COD, 80% of TN and 67% of TP were reached in an 8-hour cycle and 50% volume exchange ratio. The results were obtained after fully activated sludge granulation and consecutive biomass sedimentation. The soluble effluent COD was reduced to 125 mg/L. Industrial effluents are more difficult to treat than synthetic ones. The lower maximum OLRs also reduced the SBR granular sludge efficiency. In a bench-scale SBR, raw industrial dairy wastewater was treated with *Lactobacillus casei* TISTR 1500. Microaerobic conditions maintained in the SBR allow for biomass accumulation in large amounts, leading to 85% lactose reduction *via* rapid fermentation and subsequent protein coagulation by 90%. As a consequence, 70% of COD degradation can be achieved. Around 2.67 times higher OLR was achieved in two laboratory aerobic SBRs treated with a mixed landfill and dairy effluent than in traditional SBR processes. The best BOD_5 removal mode was reached at OLR, expressed as BOD_5, of 0.8 kg/(m³·day) per a 10- -day HRT. The application of flexible fibre as an activated sludge carrier increases the laboratory SBR reliability and it is possible to treat dairy effluents at very high OLRs. At OLR, expressed as COD, of 0.4 kg/(m³·day), COD was degraded by more than 89% and up to 97% at OLR, expressed as COD, of 2.74 kg/(m³·day). Membrane technologies are successfully applied in the treatment of low-load dairy effluents in an SBR. A high BOD removal (over 97%) and TSS-free wastewater are obtained. Due to low influent loading, TN removal reaches 96% by means of assimilation only. TP elimination reaches only 80% after system optimisation due to the limited excess sludge disposal.

Moving bed biofilm reactor (MBBR) shows very high performance when applied to dairy wastewaters: OLR increases dozens of times compared to conventional activated sludge systems. A milk processing effluent was treated in a MBBR with biomass developed on FLOCOR- -RMP particles (Henderson Plastics Ltd, Norfolk, UK). At OLR, expressed as COD, of 5 kg/(m³·day), more than 80% of total COD degradation was achieved in almost half-order kinetics with partial substrate penetration. TN was decreased by 13.3–96.2%. The small reactor volume and the high OLR encompass process applications including plant renovation and the introduction of new, limited-space treatment facilities. A novel MBBR with free-floating plastic elements (with a density slightly less than 1.0 kg/m³) may give 85 and 60% COD reduction at OLRs of 12 and 21.6 kg/(m³·day), respectively. On the basis of test results, we can say that the MBBR should be very suitable for the treatment of dairy industry effluents.

Good results can be reached in a membrane bioreactor during the treatment of an ice-cream factory effluent with 13.3 kg/m³ of COD, 6.5 kg/m³ of BOD_5 at a temperature of 25 °C. Both indicators are reduced by over 95%, while TKN is decreased by more than 96 and TP by 80%. Under aerobic conditions, the indigenous microflora composed of lactic acid bacteria may reach over 10^9 CFU/mL, which will downgrade CIP-induced alkaline pH variations.

Various alternatives for aerobic treatment of dairy effluents are also used. Pure oxygen is another possibility in the biodegradation of milk wastewater. Oxygen can be applied directly in the homogenisation tank during a traditional physicochemical treatment and stable operation is achieved under a broad initial COD and TSS range. This modification improves effluent quality and reduces process costs. Such oxygen injection systems can replace the expensive anaerobic treatment and are naturally safer. Cheese whey can also be successfully utilised as a cheap medium for edible mushroom cultivation. Some authors report the growth of *Ganoderma lucidum* on protein-free cheese whey. The best soluble COD utilisation was achieved at pH=4.6 and 27.1 °C, while the maximum mycelial yield of 0.35 mg per mg of soluble COD removed was obtained at pH=4.2 and 28.5 °C. Although there is information on edible fungal growth, dairy wastewater utilisation has not been studied from a COD point of view.

Cheese whey effluents can be treated successfully in municipal wastewater treatment plants. Factories with onsite treatment technologies should collect sanitary wastewater independently from processing effluents and discharge them directly into municipal wastewater treatment plants. Nevertheless, such a treatment option can lead to operational problems with secondary treatment units. Periodic sludge bulking is possible and is caused by intermittent high soluble COD levels in the receiving sewage plant.

Anaerobic processes: Anaerobic systems are more suitable for the direct utilisation of high-strength dairy wastewater and are more cost-effective than aerobic processes. If properly operated, these systems do not produce unpleasant odours. The major problems of anaerobic dairy wastewater treatment include long start-up periods due to complex substrate degradation, preliminary biomass adaptation prior to protein and fat utilisation, fast drop in pH and a resultant inhibition of methane production (as a consequence of the high concentration of easily fermentable lactose and low substrate alkalinity), sludge disintegration by fats in the form of triglyceride emulsions and subsequent biomass flotation, presence of inhibitory compounds (long-chain fatty acids, K^+ and Na^+ ions), inability of ammonia biodegradation and phosphorus removal, careful management, increased sensitivity to various OLRs and shock loadings, *etc.* Notwithstanding the little information on industrial-scale anaerobic plants utilising cheese whey, more than 75% COD removal and around 10 kg/(m³·day) of OLRs, expressed as COD, are achieved. The degree of biodegradation depends on the HRT applied.

Milk processing effluents are predominantly treated in conventional one-phase systems: upflow anaerobic sludge blanket (UASB) reactor and anaerobic filter (AF) are most commonly applied. UASB reactors have been used in industrial dairy wastewater treatment for more than 20 years. They are suitable for treatment of overloaded effluents with COD higher than 42 g/L. Laboratory scale UASB reactors utilising whey permeates in a continuous regime have been designed. Kinetic coefficients using the Monod equation are determined per HRT of 0.4–5 days and an initial wastewater COD of 10.4–0.2 g/L. It was shown by a comparative study of the possibility of using flocculent sludge and the effect of different HRTs (6–16 h) on the anaerobic UASB reactor behaviour applied to dairy wastewater treatment that nearly 80% of protein mineralisation, soluble COD and volatile fatty acid degradation as well as over 60% fat removal can be reached at an HRT of at least 12 h and an OLR, expressed as COD, of less than 2.5 g/(L·day). Biomass granulation was also achieved in the UASB reactor within 60–70 days. Of all the elements studied, only Ca^{2+} ions had any significant effect. When treating a synthetic ice-cream effluent in the UASB reactor, TOC was reduced by 86% at an HRT of 18.4 h, with the highest OLR, expressed as TOC, reaching 3.06 kg/(m^3·day). High FOG degradation is also possible in an UASB reactor. A couple of bench-scale UASB reactors were successfully employed during the utilisation of a synthetic milk effluent rich in FOG (0.2, 0.6 and 1 g/L). Enzymatic pre-hydrolysis contributed to 8% more COD removal at the highest FOG concentration. Cheese effluents are degraded in UASB reactors in laboratory tests and on an industrial scale. A laboratory-scale UASB reactor utilising a cheese factory effluent eliminates around 90% of effluents at an OLR, expressed as COD, of 31 g/(L·day). Organic loads, expressed as COD, over 45 g/(L·day) perform worse (70–80% only). Moreover, chemicals are needed to support a constant pH. Short-shock OLR during operation increases sludge granulation, improving stability in reactor performance. The results of the laboratory tests on an industrial level have been confirmed, improving them by 6% per 10% higher load. A full-plant UASB reactor can be applied in cheese factory wastewater treatment. With an initial COD of 33 g/L, HRT of 16 h and OLR, expressed as COD, of 49.5 kg/(m^3·day), 86% degradation can be reached. During the utilisation of an industrial effluent from Edam cheese, butter and milk production, a full-scale UASB reactor can be applied, the COD being decreased by 70%.

Dairy effluents with a low TSS can be successfully utilised in AFs in an all-scale range. The COD decreased by between 60 and 98% at a HRT of 12–48 h and an OLR, expressed as COD, of 1.7–20 kg/(m^3·day). A large specific surface of the filter media creates a precondition for higher biomass accumulation which is less affected by shear stress. A five-time higher load than with the non- -porous filler under the same conditions is achieved. It has been reported that with a couple of mesophilic upflow AFs utilising a milk bottling effluent, the reactor with the porous packing performed better (OLR, expressed as COD, of 21 kg/(m^3·day)) than the same reactor with non-porous packing (OLR, expressed as COD, of 4 kg/(m^3·day)), which is influenced by shear stress to a greater extent. Different temperature regimes can be analysed during the treatment of dairy wastewater in laboratory upflow AFs. At 12.5, 21 and 30 °C and HRT of 4 days on average, the COD removal in each reactor amounted to 92, 85 and 78%, respectively. An AF was used to treat ice-cream wastewater in a comparative study with contact process, UASB reactor and fluidised bed bioreactor (FBB). The data showed a COD removal of 67, 80 and 50% at OLR, expressed as COD, of 6, 1 and 2 kg/(m^3·-day) and 60% of total COD removal, at OLR, expressed as COD, of 2–4 kg/(m^3·day). All reactors had a poor biomass retention resulting from FOG loading. An upflow AF performed better, which allowed its full-scale installation in the manufacturing process. An upflow AF has been claimed to be unsuitable for the anaerobic digestion of very dilute dairy wastewaters. In fact, continuous stirred-tank (CSTR), UASB and baffled reactors also cause problems although experimental data show that the baffled reactor performs better with an OLR, expressed as VS, of 0.117–1.303 g/(L·-day) and HRTs between 18.8 and 2 days.

Although a CSTR is a good option for scientific research of complete-mix systems, it is difficult to use it on an industrial scale because of HRT restrictions. Such reactors were studied with a cheese effluent consisting of wash water/whey ratio of 4:1 with 17 g/L of COD. However, problems with sludge loss arise if the HRT drops to below 9 days.

Milk processing effluents can be treated in hybrid systems too. An anaerobic contact digester may reach a COD degradation of over 80–95% under mesophilic conditions. The main disadvantage is the difficult sludge settlement. However, the technology is applied worldwide in dairy plants although it is quite old. A laboratory-scale experiment analysed the kinetic performance of anaerobic synthetic ice-cream effluent at 37 °C applying the Monod and Contois equations at an HRT range between 2.99 and 7.45 days. A better explanation of the kinetic coefficients can be achieved in the final pilot-scale plant since it allows variations in the initial substrate concentration.

Anaerobic packed-bed bioreactor (PBB) can be successfully applied for dairy wastewater treatment of various organic loads. A downflow PBB was used for treating deproteinised cheese whey with 59 g/L of COD. At OLR, expressed as COD, of 12.5 kg/(m³·day), the system decreased the COD to 90–95% at HRT of 2–2.5 days. The influent pH was around 2.9, while the pH in the reactor was almost neutral. Good results were obtained in a pilot--scale plant with an up-flow anaerobic PBB. The initial cheese whey COD was 59.4 g/L. A 16-hour HRT was enough to reach 99.4% of lactose conversion. Whey wastewater was degraded to 89% in an anaerobic MBBR at (35±2) °C per 1-day HRT and an OLR, expressed as COD, of 11.6 kg/(m³·day). The cheese whey was decomposed in a laboratory PBB with a polyethylene carrier. The highest COD reduction was achieved at a 3.5-day HRT with OLR, expressed as COD, of 3.8 kg/(m³·day) and biogas production of 0.42 m³ per kg of COD per day. The mesophilic anaerobic fluidized-bed bioreactor system degraded 5.2 g/L of COD in the ice-cream wastewater to 94.4% at 35 °C, OLR, expressed as COD, of 15.6 kg/(m³·day) and HRT of 8 h. Under shock loading, the return to steady-state conditions was possible within 6–16 h. The fluidized-bed bioreactor was used to treat a low-load milk effluent with 0.2–0.5 g/L of COD. At an 8-hour HRT, 80% of COD was removed.

Membrane applications in anaerobic systems are good options for improved effluent filtration combined with a higher concentration and an effective differentiation between HRT and solids retention time. A completely mixed anaerobic microfiltration membrane reactor system was used on cheese whey high in COD (63 g/L). More than 99% of organic matter was utilised when HRT was 7.5 days, which allowed authors to upgrade the studies from the pilot plant to a full-scale demonstration. The application of the ultrafiltration system made it possible to achieve a higher biomass retention for more efficient wastewater treatment.

Different temperature conditions have been tested in order to reach a higher COD anaerobic removal. The psychrophilic anaerobic operation in some laboratory hybrid reactors, utilising whey effluents with low (COD of 1 kg/m³) and high (COD of 10 kg/m³) load, showed a better COD performance when the OLR reached 70–80% in the first reactor (at OLRs, expressed as COD, of 0.5–1.3 kg/(m³·day), in a 20–12 °C range) and more than 90% in the second (at OLRs, expressed as COD, up to 13.3 kg/(m³·day), in a 20–14 °C range). If the high-load reactor was operated at 12 °C, COD removal decreased to 50–60% and biogranule decomposition started. These side effects could be eliminated *via* an OLR reduction down to 6.6 kg/(m³·day). However, dairy wastewater has higher average temperature, which makes it possible to apply high-load wastewater treatment technologies. Another study showed that mesophilic conditions ((36±1) °C) generate more H_2 compared to thermophilic ones ((55±1) °C) during the treatment of cheese whey wastewater, with 9.2 and 8.1 mmol of H_2 per g of COD, respectively. The specific H_2 production was 4.6 times higher at 36 than at 55 °C.

Separated-phase systems are preferred from technological point of view. They have the highest organic loading and shortest HRT compared to other anaerobic digesters. The consecutive acidogenic-methanogenic phase division of anaerobic digestion is suitable for the treatment of dairy wastewater with an unbalanced composition (high C:N ratios which acidify very quickly). In such separated-phase systems, the acidogenic reactor has a major role as it supplies short-chain volatile fatty acids which can be easily fermented to CH_4 in the methanogenic reactor. The easily utilisable lactose requires a shorter HRT and a smaller volume of the acidogenic reactor than the methanogenic digester. Such a system was used to treat a dairy effluent with 50 kg/m³ of COD and pH=4.5. The COD was decreased by 72% at 35 °C and the following operating conditions: OLR, expressed as COD, of 50 and 9 kg/(m³·day), when HRT was only 1 and 3.3 days in the acidogenic and the methanogenic reactors, respectively. The CSTR was the preferable model for the acidogenic phase. In a 9-month operation study, a two- -phase anaerobic reactor comprising an acidogenic-phase CSTR and a methanogenic-phase upflow AF was used to treat dairy waste streams. The effluent COD was reduced by 90% and the BOD_5 by 95%, while an OLR, expressed as COD, of 5 kg/(m³·day) and a 2-day HRT were obtained. The H_2 and subsequent CH_4 production from fresh cheese whey were achieved in a CSTR, at 35 °C and HRT of 1 day. The mixed liquor was consequently fermented to CH_4 in a baffled bioreactor, operated at HRTs of 20, 10 and 4.4 days. At the lowest HRT, the COD reduction reached 94%. An acidogenic CSTR and a final methanogenic upflow AF were used to utilise cheese whey. The results showed that a maximum acidogenesis of up to 50%, with the same OLR (expressed as COD) range (0.5–2 g per mixed liquor suspended solids per day) could be achieved at an HRT of 24 h. The effluent was fed subsequently to the upflow AF where the initial soluble COD was decreased by 90% during HRT of 4 days. A two-stage hybrid UASB reactor, filled respectively with polyurethane foam and polyvinyl chloride rings in each phase, was supposed to exceed other anaerobic methods in the treatment of dairy effluents. The combined COD removal in the reactor in a stable equilibrium (10.7 to 19.2 kg/(m³·day)) changed from 97 to 99%. Carrier incorporation into anaerobic reactors for biomass support greatly increases their specific activity. Depending on the operating temperature, dairy wastewater can be treated in a two-phase separation. The basic configuration presupposes that thermophilic acidogenesis is followed by mesophilic methanogenesis. The information on these processes in the literature is scarce. An experiment compared two couples of anaerobic SBRs working at the following temperatures: the first couple (thermophilic-mesophilic system) at 55–35 °C and the second (mesophilic-mesophilic system) at 35–35 °C. At an OLR, expressed as VS, varying between 2–4 g/(L·day), the thermophilic-mesophilic system performs better (VS removal rate of 43.8–44.1% when HRT is 3 days and 37.1–38.9% when HRT is 6 days) than the mesophilic-mesophilic system (VS removal rate of 29.3– 30.2% when HRT is 3 days and 26.1–29.1% when HRT is 6 days). The overall improved performance showed that the thermophilic-mesophilic system with respect to total coliform reduction, TSS removal and biogas production, is preferable to the mesophilic-mesophilic SBR couple. Despite that, higher energy consumption during the thermophilic phase should be taken into account from an economical point of view. During a set of experiments, a high-temperature-based technology including acetic and butyric acid fermentation followed by CH_4 production achieved 116% COD reduction and 43% CH_4 biosynthesis, thus performing better than single-phased processes.

Combined (anaerobic-aerobic) processes. Since an anaerobic technology reduces mostly C-containing contaminants and has a weaker effect on nutrient removal, it needs to be considered as only a preliminary step which must be polished. This can be achieved by incorporating a local aerobic step or, occasionally, by directly discharging anaerobic effluent into the municipal wastewater treatment plants.

A mixed dairy wastewater was purified on a full-scale level in consecutive UASB reactor and aerobic denitrification steps. When 95% COD removal was achieved, the produced CH_4 was sufficient to cover the plant energy requirements.

SBR great flexibility makes it an adequate post-aerobic step in combined dairy wastewater treatment. A new downflow-upflow hybrid reactor containing downflow pre-acidification and upflow methanation chambers was designed to treat high-load cheese wastewater at an average OLR, expressed as COD, of 10 g/(L·day). COD (98%) was converted into biogas, while the discharged soluble COD reached 1 g/L. The process was maintained at stable pH values without chemical addition. After treatment in the SBR, more than 90% of COD, nitrogen from NH_3 and TP were removed. Wastewaters from raw milk quality laboratories, containing milk preservatives (sodium azide or chloramphenicol), were utilised in an industrial-scale plant with an AF and SBR. Influent FOG were completely treated in the anaerobic step without biomass washout for more than 2 years of operation, the COD decrease being more than 90% at an OLR, expressed as COD, of 5–6 kg/(m³·day). However, alkali had to be added to reduce the critical pH drop. The outgoing stream from the anaerobic process was polished in SBR until the final COD dropped to 200 mg/L and the TN to less than 10 mg/L.

The consecutive anaerobic-aerobic technology was used to purify reconstituted whey wastewater in a single reactor at low oxygen concentration and 20 °C. Maximum COD removal of (98±2) % was reached at total cycle time of 4 days and OLR, expressed as COD, of 0.78 g/(L·day). In accordance with specific biomass activity, trophic differentiation can be seen in the system: methanogens predominantly live at the bottom of the bulk liquid, while acidogens inhabit suspended flocs. When the soluble O_2 rose to 0.5 mg/L during the aerobic phase, the COD was reduced to (88±3) % in a 2-day total cycle time at 1.55 kg/(m³·day).

The discontinuous manufacturing process and high production heterogeneity in milk processing make it hard to outline the general dairy wastewater characteristics. Nevertheless, it can be concluded that dairy factories are large water consumers and therefore produce unstable waste streams with increased temperatures, variable pH values, high COD, BOD, FOG, N and P concentrations in combination with inhibiting cleaning agents and strong fluctuations in all factors. However, there is little information on the composition of wastewater streams from certain dairy industry branches, such as the production of yoghurt and whey products, which require more attention in future research.

Conventional aerobic activated sludge systems and percolating filters are not appropriate for dairy wastewater treatment. The high soluble COD values in wastewater account for the vast filamentous growth, which obstructs proper treatment and plant management. The application of immobilised biofilm technologies offers the opportunity to treat concentrated wastewater. MBBR are promising systems. However, many studies should be performed on other dairy wastewater streams, such as high FOG effluents, acid whey, etc.

High organic contamination levels create conditions for the preference of anaerobic digestion over aerobic processes in dairy wastewater utilisation although anaerobic treatment rarely produces clear streams. This necessitates the development of novel, more effective fermentation technologies to deal with high-strength dairy effluents. Insufficient information on temperature-phased anaerobic biodegradation paves the way for new research on dairy wastewater management. A major problem in the anaerobic fermentation of dairy wastewater is ammonia, known for its toxicity if generated in high concentrations. Research can contribute a lot to the anaerobic ammonium oxidation application in the treatment of anaerobic effluents from dairy manufacturing for an improved nitrogen removal.

The consecutive combination of fermentative and oxygen processes may be a solution for appropriate milk processing wastewater treatment. However, innovative and more compact equipment should be designed to meet the challenges associated with wastewater treatment limitations and water-quality requirements. Moreover, the replacement of outdated equipment with new machines needs to be supported by more, real-case studies, which will help us understand better dairy wastewater treatment.

Treatment of Wastewater from Cement and Ceramic Industry

Portland cement is a powder that, when mixed with water, will bind sand and stone into a hardened mass called concrete. Portland cement concrete is an attractive construction product due to its low cost, high compressive strength and durability.

Concrete is an important ingredient in economic development, especially the development of infrastructure and large public projects for the development of natural or human resources, such as dams, bridges, railroads, schools, airports and the like.

A ceramic is an inorganic, nonmetallic solid prepared by the action of heat and subsequent cooling. Ceramic materials may have a crystalline or partly crystalline structure or may be amorphous (e.g. a glass). Because most common ceramics are crystalline, the definition of ceramic is often restricted to inorganic crystalline materials, as opposed to the non-crystalline glasses.

The earliest ceramics were pottery objects made from clay, either by itself or mixed with other materials, hardened in fire. Later ceramics were glazed and fired to create a coloured, smooth surface. Ceramics now include domestic, industrial and building products and art objects.

Cement Industry

The basic raw material for the production of cement is lime. Lime is obtained from a variety of sources, primarily limestone, cement rock, oyster shell marl or chalk, all of which are primarily calcium carbonate. In addition, silica, alumina and iron ore are needed. These are obtained from sand, clay, shale, iron ore and blast-furnace slag.

The selection and amount of additional ingredients is a function of the desired properties of the cement produced. The raw material grinding and kiln steps can be performed by either of two production processes, wet or dry. The choice between the two depends on the water and chemical content of the raw-materials, the availability of process water and the cost and availability of fuel.

In the wet process the raw materials are ground with water and are fed into the kiln as slurry. In the dry process the raw materials are dried, dry ground and fed into the kiln pneumatically. The remaining steps are identical in the wet and dry processes.

Water use

The cement industry in terms of total water use does play an important role in most industrial economics. Since the cement industry is found in so many geographical locations the water use characteristics vary quite a bit from country to country.

Water in the cement production process has three basic uses:

Cooling Water

The major water use at most cement plants is for cooling. Cooling water is used for bearings on the kiln and grinding equipment, air compressors, burner pipes and finished cement. Most cooling is non-contact. Cooling use is approximately the same for both the wet and dry processes.

Process Water

Process water is needed only for the wet process. Here water is added to the raw materials to aid in grinding and to make slurry for feeding the ground material to the kiln. The process water enters the kiln as part of the slurry and is evaporated, providing no liquid wastes.

Service and Sanitary Water

Water is needed in certain plants for the preparation of raw material, either washing or beneficiation. Water is also used in the disposal of collected kiln dust. In the wet process, kiln dust is leached of soluble alkalis to recover raw materials by mixing the dry dust with water to form slurry.

The slurry is then put into a clarifier. The treated dust slurry is returned to be used in the raw material slurry preparation. For both the wet and dry processes, collected dust can be mixed with water and fed to a settling pond where the settled solids are not reused and the clarified water discharged.

Although this method is practised, it is not recommended to discharge untreated Wastewater. Large quantities of cement dust are generated due to grinding and handling of cement. Water is used to control cement dust by being sprayed on roadways and parking lots and used for washing trucks.

One possible use for water in the cement industry is for the prevention of air pollution. This use is not frequently employed now, but may become more important in the future. The air pollution from the kiln stacks can be treated by wet gas-scrubbers. The scrubbers will use large quantities of water (on the order of 10 times that of the total water use). Therefore, a former air pollution problem becomes a water pollution problem and the scrubber effluent must be treated.

Recycling and Reuse

There is some potential for recycling and reuse of Wastewater within the cement production process. Cooling water can easily be recycled by installing cooling towers or cooling ponds for the dissipation of waste heat.

For the wet process, cooling towers or pond blow-down can be used in raw material slurry preparation. In a wet or dry-process plant, if kiln-dust leaching is being used, the blow-down can be used in the preparation of kiln-dust slurry. In a wet-process plant, all other Wastewater can be reused in the kiln process except the leachate effluent from the clarifier, which must be disposed of in a containment pond. In the dry process, there are limited numbers of reuse possibilities. However, leachate waste could be treated to produce water suitable for reuse by means of electrodialysis.

Sources and Characteristics of Wastewater

Similar wastes are produced by the wet and dry process. However, in the wet process the wastes enter the Wastewater stream while in the dry process the wastes enter the atmosphere. In the wet process, spillage and overflow from slurry formation produce suspended solids, dissolved solids and alkalinity in the preparation and grinding stage.

However, the major sources of Wastewaters are generally produced when addressing the dust or air pollution problem, i.e. the collection of kiln dust and its disposal. The amount of waste generated by the wet process is slightly more than the dry process due to more use of leaching.

Wastewater Treatment Practices

Few steps in the manufacture of cement directly produce liquid wastes. In non-leaching plants, contact of raw material or final product with water provides the major source of the waste load. These waste sources can be reduced through good cleaning and maintenance practices or by collecting these waste sources for treatment. One important area of waste is the storage of raw material and finished products.

Protection of these materials from precipitation and spillage on the plant grounds will reduce significantly pollutants entering the Wastewater stream. For leaching plants, settling ponds, containment ponds and clarifying are used as pre-treatment processes.

The discharge of the containment pond is then treated by electrodialysis for final discharge or recycled for slurry formation. The cooling water can be recycled through cooling towers or ponds and the blow-down reused.

Ceramic Industry

Manufacture of ceramic is an ancient art. In general ceramic can be defined as 'the art and science of making solid articles by the action of heat on earthen materials, which have inorganic nonmetallic materials as their essential component'.

This definition includes not only materials, such as pottery, porcelain, refractories, structural clay products, abrasives, porcelain enamels, cement and glass but also non-metallic magnetic materials; ferrotrics manufactured single crystals, glass ceramic and variety of products which were not in existence a few years ago. Ceramic products are wide ranging.

Each of these products need different composition of raw materials different glazing materials and also to be fired to a definite maximum temperature, which differ for each product, and to a definite firing and cooling time temperature schedule.

Thus manufacturing processes for production of ceramics are equally wide ranging. However, the basic manufacturing process remains same. In general the manufacturing process can be divided in following steps.

Recommended wastewater treatment options for units manufacturing various ceramic products:

> Due to different nature of industrial (non-biodegradable) and domestic (biodegradable) Wastewater from such units, segregation of the two is recommended.

After the initial step of segregation, the following treatment systems are recommended for these Wastewaters:

For Industrial Wastewater

The Wastewater from ball mill washing and propylene drum washing should be taken to independent settling tanks. The settled sludge should be periodically removed and dispose of to commensurate with applicable disposal practices, because the sludge from glaze preparation section would contain heavy metals and thus need to be disposed of in a secure manner.

The overflow from these settling tanks should be taken to the effluent treatment plant for final treatment. Such a system would help in reducing the load on effluent treatment plant.

Wastewater Characteristics

The results of the samples collected at the inlet and outlet of settling tank are presented in Table.

Table: Characteristics of Wastewater at inlet and outlet of setting tank.

Source	pH	TSS mg/l	BOD mg/l	COD mg/l	Oil and Grease mg/l
Inlet	7.8	1817	<10	840	12
Outlet	8.0	163	<10	<15	12

The characteristics of the effluent show that the unit can dispose its water in municipal sewer. However, in the absence of any such facility, physico-chemical treatment of Wastewater to remove high TSS and O/G concentration is required before its final disposal.

Sources of Wastewater Generation

The sources of Wastewater generation in the unit are as follows:

Industrial

About 52.5 m³/day (estimated) of industrial Wastewater is discharged to the municipal sewage every day. Major industrial usage of raw water is in the pickling section and in ball mills section.

Total discharge of Wastewater from the pickling section is estimated to be of the order of 22.5 Kl/day. The degreasing tank is reported to be cleaned after every 3-4 months and the effluent is utilised in manufacturing scouring powder (used for washing of utensils) by mixing with rice husk ash and sold to local market.

The pickling and neutralisation tank are cleaned every 15 days and 7 days respectively and the Wastewater from both the tanks flow to a collection sump. Thus neutralisation takes place in the collection sump in course of time and the neutralised effluent is discharged to the municipal drain. Expected pollution parameters in the Wastewater are pH TSS, and grease and Fe.

Domestic

Domestic water is used in the toilets and for drinking purpose. It is reported that about 18 m³/day of domestic Wastewater is discharged to the municipal sewer.

Wastewater treatment system: The unit does not have any Wastewater treatment system either for industrial or for domestic Wastewater, and discharges their effluent Wastewater (of the order of 45 m³/day) to the municipal sewer.

Wastewater characteristics- Spot samples of water have been collected from the pickling unit and the ball mill section and analysed in government approved laboratories, for pH, TSS, oil and greases, metals, results of which are summarised in table.

Table: Wastewater characteristics from different sources.

Source of sample	BOD	pH	TSS mg/l	CO	Ni	Fe	Heavy metals mg/l Na	Ca	K	Oil and grease mg/l
Post alkali treatment rinsing	<30	7.2	114	-	-	-	-	-	-	37
Post acid treatment rinsing	<30	7.1	714	-	-	2.82	-	-	-	27
Ball mill washing	<30	7.0	48	0.1	0.6	-	255	148	7	-
Final discharge	<30	7.3	81	-	-	0.88	-	-	-	-

Water Pollution Prevention Techniques

In ceramic industries water is used mainly in following areas:

* Wet grinding of raw materials.

- Preparation of slip for moulding the required shape.

- Glaze preparation.

- Glazing (in spray glazing).

- Flour washing and equipment/container washing.

- Domestic use.

The water pollution prevention techniques can be divided in two types of measures as discussed below:

Housekeeping Measures

It is a common practice that hose pipes used for floor washing, equipment washing are kept open, resulting increase in total volume to be treated in effluent treatment plant, i.e. higher capital investment and recurring cost.

A simple measure, i.e. usage of self-closing type water hose pipes will reduce the avoidable capital and recurring cost of the effluent treatment plant. Careful handling of fuel oil to prevent spillage will help in bringing down the oil and grease in Wastewater.

Process Modifications Operational

Wet grinding is done mostly in ball mills, from where the slurry is sent to underground blungers for blungering. In production of potteries, stone wares, sanitary wares, porcelain fire bricks, etc. after Hungering, the excess water is removed in filter presses. In many cases, the tank capacity does not commensurate the wash water hence part of it overflows down the drain.

Sometimes part of the water is intentionally drained for quality reasons. Efforts should be made to maximise utilisation of wash water. At least the filter press wash water could be completely recycled without any adverse effect on quality.

Prepared glaze is normally stored in PVC or metallic containers. In case of dip-glazing there is no water discharge, however in case of automatic spraying, the un-utilised glaze from spray which normally is washed down the drawing can be recycled. If not fully, at least the first wash from the drain could be recycled for use.

In ball mill and other equipment washing (in raw material handling sections), it is a common practice to discharge the whole wash water, which is quite substantial. It is recommended that, the equipment should first be rinsed with a little quantity of water and wash water should be collected separately and recycled. If process permits then subsequent wash water can also be recycled.

Otherwise it can be taken to a settling tank and overflow to the effluent treatment plant. The wash water is the major source of TSS and heavy metals. The data from one such unit shows that the TSS concentration from such washing is more than 25,000 mg/L. Avoidance of wash water will not only reduce the pollution load, but also reduce the capital and recurring cost of effluent treatment plant.

Water Pollution

The water discharged from potteries, porcelain, small scale sanitary ware manufacturing units, decoration wares, fire bricks, stone wares is very low. However, in big units and specially sanitary wares, tiles manufacturing units the water discharged is substantial.

The characteristic of the Wastewater analysis shows the presence of high TSS concentration along with heavy metal depending upon the glaze. The Wastewater should be treated before its final disposal. For proper design of Wastewater treatment system, it is recommended to segregate non-biodegrade industrial Wastewater from bio-degradable domestic Wastewater of the unit.

The segregated industrial Wastewater is then recommended to be treated by dosing alum singly or in combination with polyelectrolytes, followed by sedimentation. For domestic Wastewater anaerobic treatment in a septic tank is recommended before its final disposal. However, if sewer facilities exist the domestic water can be directly discharged into sewer without any treatment.

Coal Plant Wastewater Treatment

Flue gas desulphurization (FGD) wastewaters are produced at coal-fired power plants in increasing quantities as the regulation on air emissions is tightened worldwide.

Low cost and environmentally favourable reuse of this wastewater stream has become an important topic with the respective national and local regulatory bodies stipulating minimum treatment levels and standards.

Traditional technologies, which are otherwise used for concentration of saline streams, fail the economic and performance benchmark that needs to be met.

Carrier gas extraction (CGE) technology, which was specifically developed to handle high levels of contamination and variability in feed waters, is optimal for this application and offers the lowest cost solution within the required performance.

Source and Constituents

Coal-based power plants generate over a third of the planet's electricity. The combustion of coal in these facilities produces a flue gas that is emitted to the atmosphere.

Many power plants are required to remove SOx emissions from the flue gas using FGD systems. The leading FGD technology used globally is wet scrubbing (85 per cent of the installations in the US and 90 per cent of the installations in China).

Commonly, three kinds of scrubbers are used for wet scrubbing-venturi, packed and spray scrubbers-and entail injection of alkaline scrubbing agents into the scrubber.

Typically, the agent is limestone (i.e., calcium carbonate), quick lime and caustic soda.

For example, when limestone reacts with SOx in the reducing conditions of the absorber, sulphur dioxide (the major component of SOx) is converted into sulfite, and a slurry rich in calcium sulfite is produced. In forced oxidation FGD systems, an oxidation reactor is used to convert calcium sulfite slurry to calcium sulfate (gypsum).

From the blowdown and dewatering processes of the slurry, the FGD wastewater stream is created. The composition of coal and limestone primarily affects the composition of the wastewater. Other parameters which have a smaller effect are the type of scrubber and the dewatering system used. Coal contributes chlorides, fluorides and sulfate to the wastewater. Because of the metallurgy used in the scrubbers it is typical to purge the wastewater before the chlorides exceed 12,000 mg/L. The level to which chlorides are tolerated by the metallurgy of the scrubber determines the amount of wastewater generated. Use of better metallurgy can help reduce the amount of wastewater by not purging until a chlorides level of up to 35,000 mg/L is reached.

Coal also adds trace metals, including arsenic, mercury, selenium, boron, cadmium and zinc. Limestone could contribute iron and aluminium to the FGD wastewater.

Varying Water Quality

Defining a standard composition of FGD wastewater across different power plants is tricky because there is no consensus in the industry on where (in the process train) the sample for measuring the composition has to be collected, and the design of the process train downstream of the scrubber itself changes from one facility to another. For example, some facilities may employ primary and secondary hydro cyclones to maximize the capture of solids before gypsum dewatering.

It is also common for a plant to change coal and limestone suppliers so the wastewater constituents will change over time during operation of the FGD system. Unlike other industrial wastewater treatment fields, FGD wastewater samples for a specific plant are likely not available for testing before the plant is designed, built and commissioned. Additionally, during the operation of the coal-fired power plant, there might be periods when the plant is not run at full capacity and the SOx levels and FGD water quality and volumes can vary.

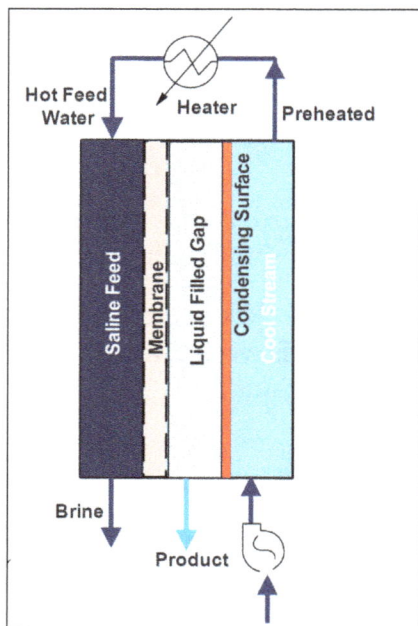

Complicating matters further, the plant's wastewater treatment system must be flexible to handle these varying inputs yet produce a treated stream that meets the plant's wastewater discharge permit requirements. All traditional treatment technologies fail because they cannot handle these requirements as they were developed for sea water or other applications where the feed water quality and volumes remain fairly constant.

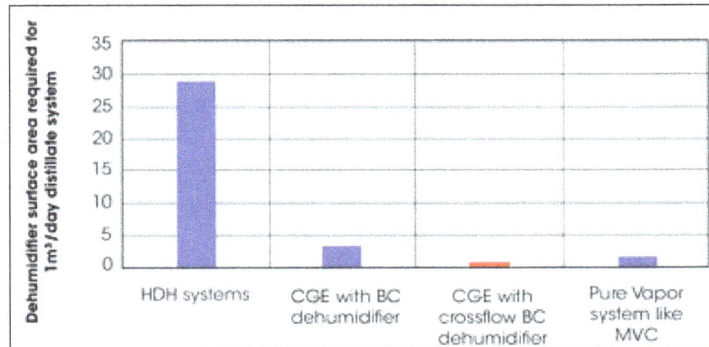

Relative condenser sizes. New CGE BC devices have 40 per cent lower surface area requirement than vapour condensers.

CGE Technology

Invented at Massachusetts Institute of Technology, CGE is a novel method of desalinating high salinity water streams using a carrier gas. The technique was specifically developed to handle high contamination wastewaters at varying volumetric rates and quality. Over 50 patent families cover various innovations which make CGE an economical technology for treating FGD wastewaters to high recovery rates and with high levels of influent water variability.

CGE is a desalination process that mimics the rain cycle. It uses a carrier gas as a medium to de-salinate saline streams by using a humidification-dehumidification configuration. CGE consists of two main unit operations: humidifier and dehumidifier.

Both the humidifier and the dehumidifier are direct contact heat and mass exchange devices. The humidifier is a packed bed device wherein a heated water stream (<90°C) is introduced in the form of droplets and the carrier gas (which is typically ambient air) is introduced at the bottom of the device in a counter-current configuration.

The carrier gas comes in direct contact with the saline droplets and there is evaporation from the surface of the droplet into the carrier gas stream. Hence, as the carrier gas rises through the device, it accumulates increasing amounts of pure water vapour from the saline stream. The concentrate, which retains all of the dissolved salts, exits the humidifier, is diluted with feed wastewater, is pre-heated with heat from the dehumidifier, is heated using a source of energy (like solar heaters) and is recirculated back into the humidifier.

The dehumidifier is a multi-stage bubble column device. In this device, the air-vapour mixture from the humidifier is sparged through several shallow layers of fresh water in successively cooler stages. As a result, small bubbles are formed and the vapour condenses from the surface of the bubbles into these shallow pools. As this fresh water is generated it also picks up the heat of condensation, which is transferred back to the feed water in the preheating heat exchanger.

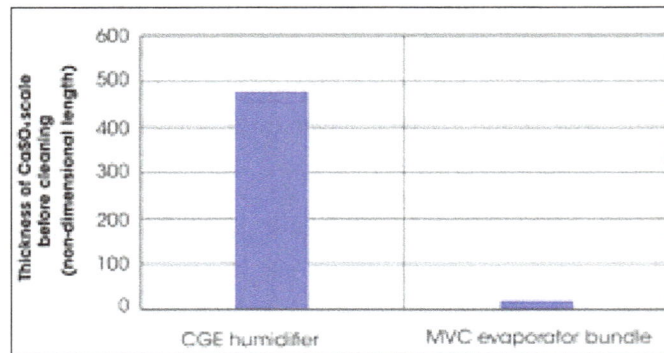

Relative scaling levels. CGE can take orders of magnitude high thickness of scale
because of the separation of the heat exchange surface from the phase change surface.

The ratio of the energy input in this preheating process to that in the heater that follows it (energy recovery ratio) has been maximized over a decade of research and development. Currently, CGE's energy recovery ratio is up to 4.5 which implies that only ~22 per cent of the total heat needed to distil the water is provided by the heat source and contributes to the energy cost. Additionally, a relatively low top brine temperature of <90°C combined with a low temperature increase required in the heater (and no steam requirement) makes the CGE optimal for using simple solar thermal energy (like evacuated tube solar heaters) as heat input, or using hot water from power plant heat recovery steam generators.

CGE was developed to handle high levels of contamination.

In addition to treating wastewater at full scale and in the lab to purity levels, the performance of CGE has also been tested with more than 100 different high contamination FGD wastewater samples provided by various industrial customers, including power plants.

The recovery of the system depends on the feed concentration as saturated brine solution is the waste/product stream leaving the system. With feed water streams of less than 120,000 ppm TDS, recovery rates are consistently greater than 95 per cent.

Treatment of industrial wastewaters is limited due to the prohibitive capital investment and operating costs associated with distillation-based techniques like mechanical vapour compression (MVC). These treatment options are expensive because:

- These technologies were not developed to handle variability in feed water quality;

- FGD wastewater is corrosive, requiring expensive corrosion-resistant metal for evaporation and condensation surfaces;

- The conventional treatment technology, i.e., MVC is not energy efficient at high salinity and the corresponding high boiling point elevation (as high as 13°C at saturation concentration of sodium chloride);

- Pretreatment requirements are much higher with high hardness levels because the evaporation and condensation surfaces are sacrosanct and cannot accumulate porous scale layers (if they do the performance of the system drops precipitously).

Since its commercialization, CGE has demonstrated a significant shift in the cost of treatment of hypersaline waters based on three main innovations.

Bubble column (BC) heat exchangers have extremely high heat and mass transfer rates as they employ direct contact condensation of the vapour-gas mixture in a column of shallow liquid unlike traditional techniques, which condense on a cold surface.

New physical understanding of heat transfer in BCs has led to low pressure-drop designs. The concept of multi-staging the uniform temperature column in several temperature steps has led to highly effective designs (about 90 per cent). These designs lead to significant cost advantages for CGE over traditional techniques like MVC distillation systems.

The colder fresh water enters the device at the topmost stage and passes through every successive stage in a cross-flow manner (from one end of the stage to the other) and extracts the heat of condensation from the condensing vapor (which heat is in turn used to preheat the feed water).

Each BC stage has a shallow layer (<one inch in height) of this fresh water through which the air/vapour mixture is sparged, creating a multitude of small bubbles. As these bubbles rise through the layer height there is a wake (negative pressure region) created below the bubble which draws in the liquid from the surrounding region. This process sets up a millimetre-size liquid circulation zone which causes continuous renewal of boundary layers formed via the heat and mass transfer phenomena. Such liquid circulations, which cause extremely high levels of turbulence, are created throughout the liquid layer because of the swarm of rising bubbles. The device is designed to be ultra-efficient with optimal bubble size, bubble pitch and liquid layer height that affects bubble residence time, gas velocity and temperature difference per stage.

The newest generation of BC devices have 40 per cent lower heat transfer area requirement than condensers in pure vapour systems like MVC. At higher salinities (as is the case with the current challenge), MVC uses Grade 5 titanium for heat transfer surfaces. The BC uses the surface of bubbles as heat and mass transfer surfaces as opposed to expensive metallic surfaces used in MVC systems, providing significant capital cost advantages due to lower heat transfer surface area requirements and the use of inexpensive materials.

Thermodynamic Balancing

When finite time thermodynamics is used to optimize the energy efficiency of thermal systems, the optimal design is one which produces the minimum entropy within the constraints of the problem (such as fixed size or cost).

This well-established principle, known generally as thermodynamic balancing, was applied to the design of combined heat and mass exchange devices (dehumidifiers and humidifiers) for improving the energy efficiency of CGE systems.

Table: Comparison of the total water production cost of CGE and MVC for 12.500 ppm feed water and 95 per cent fresh water recovery.

Project	MED	MVC	CGE
Capacity	22m3/h	20m3/h	20m3/h
CAPEX (RMB)	97.50MM	60MM	40MM
Thermal energy cost (for every tonne of influent water)	300kg	89.21kg	0 RMB/t 90° C hot water used
Electricity cost (for every tan of influent water)	30kWh	26.63 kWh	3 kWh
Total OPEX	180 RMB/m3	120 RMB/m3	31 RMB/m3

This resulted in novel designs and operating procedures that make it more energy efficient at treating hypersaline wastewater compared to traditional techniques like MVC. At the core of these innovations is a new non-dimensional parameter that was invented to minimize the average local driving force for heat and mass transfer. The fully automated optimization of the system based on this non-dimensional number gives CGE the ability to have constant performance even with varying water quality and volumes.

The physical embodiment in design of the thermodynamic balancing concept is several extraction lines through which specified amounts of the air/vapour mixture is prematurely taken from the humidifier and re-injected at a corresponding location in the dehumidifier.

In commercial CGE systems, thermodynamic states at different points in the system (at the inlet and outlet and intermediate locations of all unit operations including the humidifier, dehumidifier and heat exchangers) are determined by measuring temperatures, mass flow rates and concentrations.

Using this information, the amount of vapour/gas mixture to be extracted through any given line is determined on a continuous basis by evaluating the optimal operating point which corresponds to a non-dimensional number of 1 for any and all boundary conditions. This novel algorithm is also patented and proprietary to Gradiant.

Decoupling

The decoupling of phase change and heat transfer surfaces is crucial to treating hypersaline water streams because of hardness scaling issues. Scaling tendency increases with increasing salinity; the scale is likely to form on the surface where phase change occurs.

In a MVC system, the phase change and the heat transfer occur on the same surface, resulting in a drop in heat transfer efficiency when scale forms and acts as an insulator.

In CGE, however, the phase change occurs in the humidifier column and the feed water is heated in the heat exchanger. Scale forms on the packing material in the humidifier but does not affect the evaporation or the performance of the system because the carrier gas is in direct contact with the hypersaline water and there is no heat transfer through the material of the packing.

Nevertheless, the packing material only requires cleaning or replacement when the scale build-up severely reduces the fluid flow in the column. Another benefit of having high tolerance to scaling is that the pretreatment requirements are lower than MVC systems, which further reduces operating costs.

Economic Comparison

The CAPEX and OPEX of CGE, MED and MVC are compared for treatment of contaminated FGD wastewater from coal power plants in China for a feed water of 12,500 ppm and a fresh water recovery of 95 per cent (250,000 ppm reject stream).

The CAPEX of CGE is lower than that of MVC because of the reduced need for expensive metals like Grade 5 titanium. The OPEX of CGE is lower because of lower pretreatment requirements, use of low temperature thermal energy which can be free if solar thermal (evacuated tube) or waste heat is used, and minimal repair and maintenance needs.

In conclusion, the cost effective, high performance treatment of FGD wastewater is becoming mandatory to implement in all coal-fired power plants. Due to the variability of FGD wastewater quality and volumes on top of the high contamination levels, the use of CGE technology is optimal and is significantly more cost effective than other solutions.

Treatment of Sewage from Pulp and Paper Industry

Sewage emerging from a pulp and paper industry contains various polluting constituents because in the processes involved in pulp and paper industry various chemicals are used. As such treatment of the sewage is necessary prior to its disposal.

Processes Involved in Pulp and Paper Industry:

Raw Materials

The raw materials are:

- Cellulosic,
- Non-cellulosic.

Cellulosic Raw Materials

Bamboo is the principal cellulosic raw material. However, wood (hard or soft) is now being increasingly, used. Straw, mainly rice and wheat; grass; jute sticks; sunn hemp; old ropes; hessian; cotton linters and rags; bagasse; and waste paper are also used as raw material in small paper mills and also for specialty paper in bigger mills.

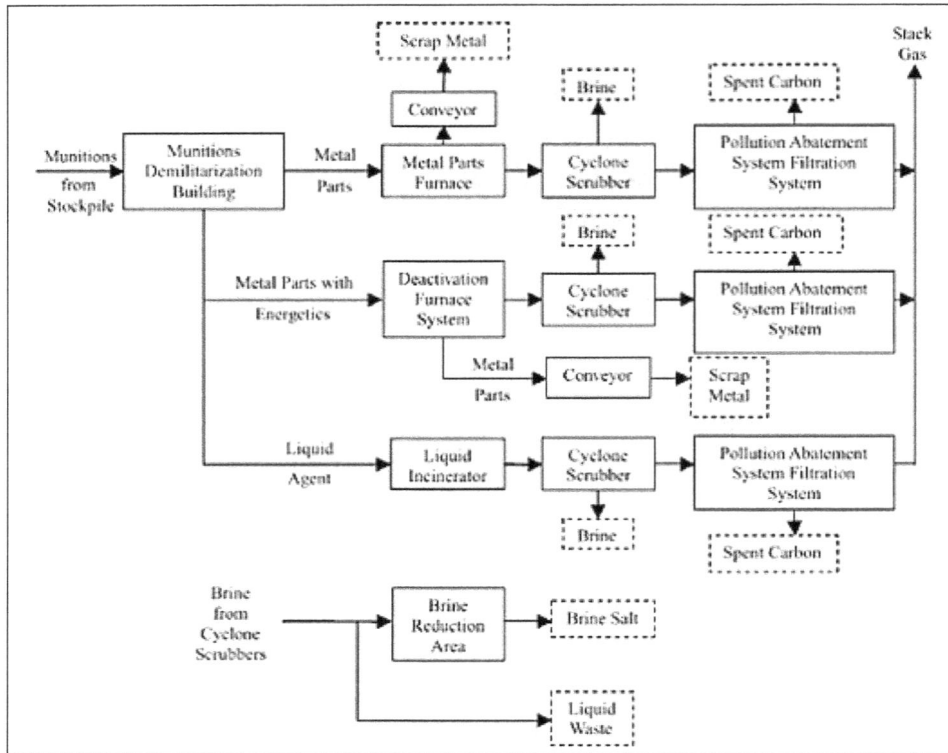

Flow diagram for the treatment of spent wash for disposal into inland surface water.

Non-cellulosic Raw Materials (Chemicals)

These are: Caustic soda, sodium sulphate, sodium sulphite, sulphur, bisulphites of calcium and magnesium, lime, limestone, chlorine, chlorine dioxide, hypochlorites of sodium and calcium, hydrogen peroxide, sodium peroxide, china clay, talc, rosin, starch, alum, glue, dyes and gums.

Preparation of Raw Material

The preparatory processes for the various types of cellulosic raw materials and pulping processes employed for them are mentioned below:

Cellulosic raw material	Preparatory process	Pulping process usually adopted
Bamboo	Dry chipping, washing, chipping, chip screening.	Sulphate, sulphite, soda, semichemical, neutral sulphite semichemical.
Wood (hard or soft)	Debarking, chipping, screening.	Sulphate, sulphite, cold soda, semichemical, neutral sulphite semichemical, mechanical.
Straw, grasses, jute sticks, hemp, old ropes, hessian, cotton linters, rags, bagasse	Chopping, dusting depithing for bagasse.	soda, sulphate, sulphite, lime soda, semichemical, neutral sulphite semichemical.
Waste paper	Sorting, dusting.	De-inking, hydro-pulping.

The pulping process also includes washing of pulp whereby the fibres are separated from other dissolved constituents either in diffusers, or vacuum or pressure filters. The separated fibres are screened to remove shieves, knots, etc., and bleached, if necessary, in several stages with chlorine, hypochlorite and other bleaching agents. One of the steps in chemical bleaching is the caustic extraction of the chlorinated pulp.

The dissolved constituents of raw materials, made up of the spent liquor and pulp washings, commonly known as black liquors, may be sent either for recovery of chemicals or byproducts, or for disposal, depending on the process adopted.

Stock Preparation and Paper Making

The making of paper from the pulp consists of two essential steps, namely – stock preparation and paper making.

Stock Preparation

In this process, the pulp is mechanically treated in beaters, refiners or other equipment to the required degree of fineness (freeness) depending on the quality of paper to be made. At this stage, sizing chemicals, loading materials, dyes, etc., are also added to impart the necessary characteristics to the paper.

Paper Making

The refined pulp is then passed on in water suspension to the paper or board machines when the sheet is formed either on continuous running wire or on the moulds. The sheets are pressed and dried.

Sources, Quantity and Characteristics of Sewage

Sewage is contributed from different sections of the pulp and paper industry as indicated below:

Raw Material Preparation Section (Chipper House)

The effluent from this section results from washing, cleaning, barking and chipping of the cellulosic raw materials.

Pulp Mill

The effluent from pulp mill consists of:

- The spent liquor known as Black Liquor (BL).
- Effluents from brown stock washers, chlorination, caustic extraction and hypochlorite bleaching.
- Chemical recovery process.
- Spills and leakages.
- Wash water from bleach liquor and chemical preparation plants.

Paper Machines

Stock preparation and paper machine effluents which include excess white drainage.

Caustic-chlorine Plant

The effluent from caustic-chlorine plant consists of:

- The sludges from brine purification and filtration and caustic filtration.
- Condensates from vacuum dechlorinators and chlorine system.
- Condensate from hydrogen cooling system.
- Other effluents from cell room and floor washes.

Quantity

The quantity of sewage produced from a pulp and paper industry depends on the total production, the types of paper made, water supply and on the practices adopted for reuse of effluents within the mill operations. As such the quantity of sewage considerably differs from mill to mill, and it may vary from about 200 to 350 m3 per tonne of paper produced.

Characteristics

The sulphite process of pulping is, however, not being used now due to great pollution exerted by the spent liquor produced in this process.

Pollutional Effects of Sewage from Pulp and Paper Industry

The main polluting constituents in the sewage from pulp and paper industry are suspended solids, colour, foam, inorganic materials such as sodium carbonate (when recovery system is not practised), toxic chemicals such as mercaptans, and inorganic sulphides. Mercury is present if mercury cells form a part of pulp and paper mill.

The sewage has high BOD and COD and when discharged untreated will damage the receiving water course due to high oxygen demanding organic and inorganic materials present in the sewage. Further the sewage imparts colour to the stream due to lignin and its derivatives present in the sewage. The colour persists for a long distance since lignin and its derivatives are not readily biologically degraded. The sewage may also impart odour to the receiving stream.

Methods of Treatment of Sewage from Pulp and Paper Industry

The treatment of sewage from pulp and paper industry may consist of the following:

- Segregation of effluents from different sections.
- Primary Treatment.
- Secondary Treatment.

Segregation of Effluents from Different Sections

Segregation of effluents from different sections of pulp and paper industry according to their characteristics facilitates in devising treatment methods at source and reducing the pollution loads in the combined effluent requiring treatment.

It also makes possible recovery and recycling of valuable materials like fibres and water. Separate collection of lignin-bearing coloured effluents will provide flexibility for colour removal for a small-volume concentrated effluent rather than the dilute total effluent.

Primary Treatment

This involves removal of suspended solids and partial reduction in BOD and other constituents in the sewage. The primary treatment methods includes coagulation, sedimentation, save-all filtration and floatation. The primary sludge along with excess secondary biological sludge may be thickened and dewatered by any of the mechanical methods or lagooned or filtered on sludge drying beds. The dried sludge may be disposed of by incineration or land-fill.

Secondary Treatment

The secondary treatment methods that may be adopted includes:

- Lagooning.

- Aerobic biological treatment including activated sludge process, trickling filters, aerated lagoons, oxidation ponds.

- Anaerobic lagoon followed by aerobic stabilization pond.

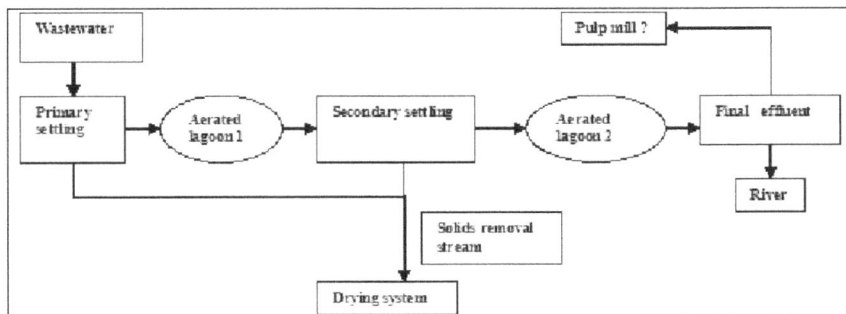

Flow Diagram for Treatment of Effluents from Different Sections of Pulp and Paper Industry.

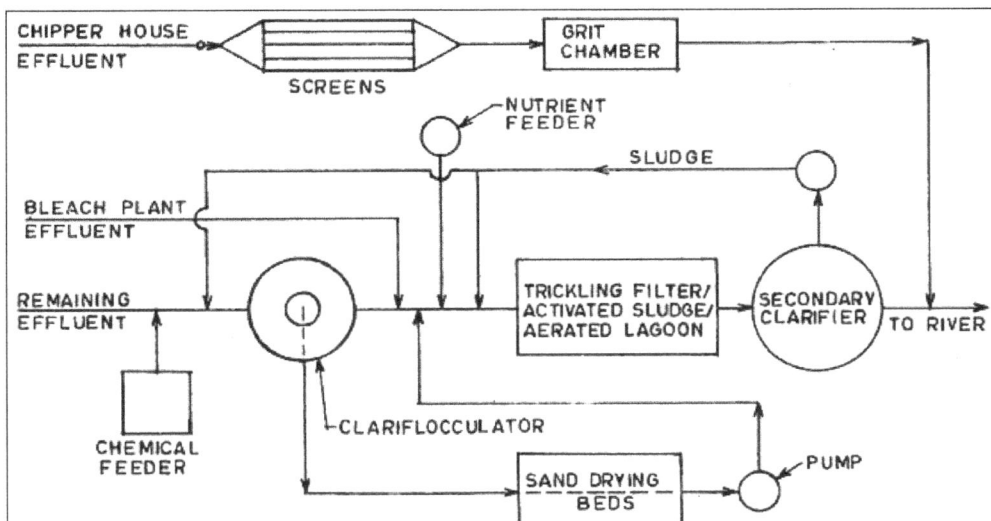

Colour bearing effluent, after suitable conditioning, may be used on land for irrigation wherever sufficient and suitable farm is available, since this method not only reduces the cost of treatment but also overcomes the problem of colour in the combined sewage discharged into river.

Treatment of Sewage from Cane Sugar Industry

Large volume of sewage is produced during manufacture of cane sugar and it contains a high pollution load. The quantity of sewage produced from a cane sugar industry and its characteristics vary widely depending on local conditions and methods of plant operations.

Process Involved in Manufacture of Cane Sugar

The various processes involved in cane sugar manufacturing can be categorized into:

- Physical processes - crushing, evaporation,
- Chemical processes - clarification.

Crushing

The sugarcane is harvested manually or mechanically. The manual harvesting eliminates much of the trash and soil from sugarcane, which reaches the factory in a condition in which it can be straightaway put through the extraction process. However, in recent times mechanical harvesting is becoming popular due to shortage of labour. Where mechanical harvesting is employed cane washing is necessary to remove extraneous material.

The cane is cut into short lengths in a shredder and then crushed in a series of roller mills to squeeze out the juice. The partially crushed cane is wetted with water in the later stages of crushing operation to aid maximum extraction. This operation is known as maceration. The extracted juice is screened to remove coarse particles of cane or bagasse and then sent to process.

Clarification of Juice

The screened juice is treated with lime to raise pH to 7.6 to 7.8 and then heated to 102°C, when coagulation of colloidal and other suspended impurities takes place. Bulk of the colour of the juice is also removed by treatment with lime, carbon dioxide is often passed through the juice for controlling its pH value.

The flocs formed in this process are settled and removed by passing the juice through a clarifier. The clarifier sludge is filtered in a vacuum filter or filter press; press mud is disposed of as solid waste or in the form of slurry and the filtrate is recycled to the process. In the sulphitation process, juice is further heated by passing sulphur dioxide gas through it when the colour of juice is bleached.

Evaporation

The clarified juice is preheated by passing it through heat exchangers and then evaporated in multiple effect evaporators and fed to a single effect evaporator for further concentration. The later process is known as 'pan boiling' in which syrup is thickened to a point where numerous small crystals of sugar are formed.

The syrup known as massecuite is then taken to crystallizers where complete crystallization of sugar takes place. The massecuite is then fed to centrifugal baskets where the sugar crystals are separated from mother liquor which drains away as molasses. During the process of centrifugation water is sprayed on the surface of sugar crystals to remove the adhering molasses.

The weak molasses thus obtained are recycled to the process. From the centrifugal baskets sugar is taken out and dried in a drier and bagged. A flow diagram of the process of manufacturing cane sugar indicating the sources of sewage is shown in Figure below.

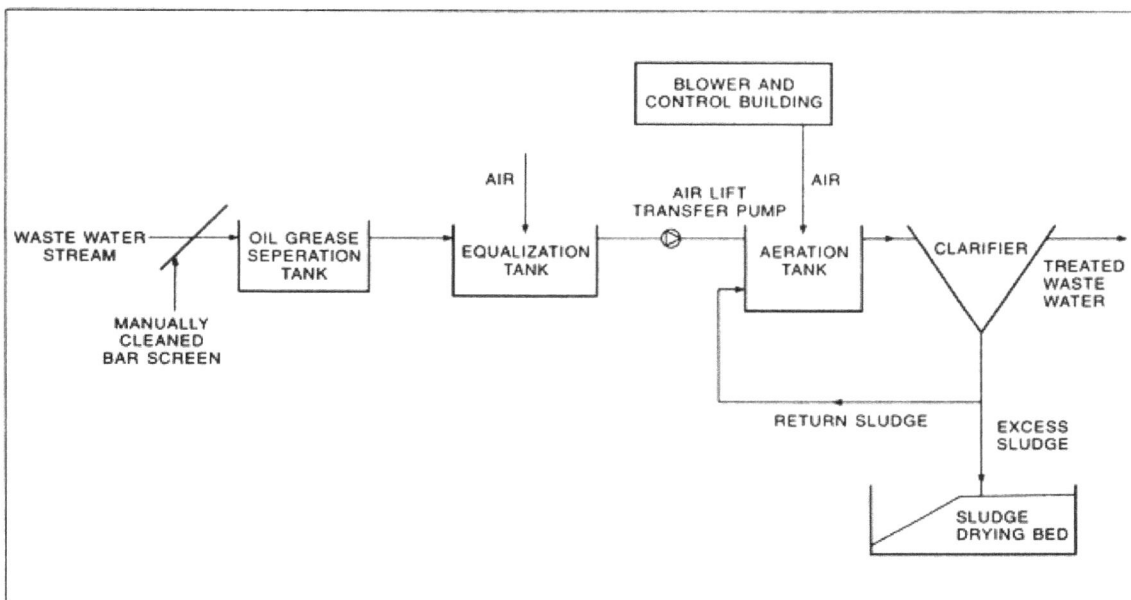

Flow Diagram of Cane Sugar Manufacture indicating Various Sections Contributing Sewage.

Sources, Quantity and Characteristics of Sewage

Sewage is contributed from different sections of cane sugar industry as indicated below:

Cane Wash Water

Cane washing is necessary where mechanical harvesting is employed. These waters contain high concentration of suspended solids which are stabilized in colloidal from by organic compounds acting as protective colloids. Cane wash waters may also contain substantial concentrations of sugar, because of damage to cane during mechanical harvesting.

Effluent from Mill House

Large amount of water is used for extraction of juice and cooling the bearings of mill tandems. This cooling water picks up oils and grease as well as some juice from spillover and leakage.

Effluent from Filter Cloth Washing

The sludge from clarifier after clarification is filtered in filter presses. The filter clothes are washed periodically, which enhances the suspended solids concentration and biochemical oxygen demand of the sewage.

Effluent from Boiling House and Floor Washing

The effluents from boiling house results from leakage from pumps, pipes, evaporators, crystallizers and especially from centrifuges together with periodic washing of floor. Although the discharge is intermittent and not large in volume, it represents the most polluting fraction of effluents because of its high biochemical oxygen demand.

The juice heaters usually foul very quickly, so it is necessary to clean heating surfaces throughout the plant to remove accumulation of scales. This is accomplished by boiling with dilute caustic soda solution and inhibited dilute hydrochloric acid followed by rinsing with water.

The sewage from boiler blow-off is of intermittent nature, but contains relatively high solids and alkalinity and low biochemical oxygen demand.

Condenser Water

In evaporators large quantity of condenser water is produced which constitute a large proportion of sewage from cane sugar industry. In the evaporation and crystallization process sugar particles gain access to cooling and condenser waters by entrainment in evaporators and vacuum pumps. This is major source of organic pollution in these waters.

Quantity

The average quantity of sewage produced from a cane sugar industry is of the order of about 3000 litres per tonne of cane crushed. Table shows the average quantity of sewage produced from the different sections of the cane sugar industry.

Table: Quantity of Sewage Produced from Different Sections of Cane Sugar Industry.

S.No	Source	Average quantity (litres/tonne cone crushed)
1	Mil house	730
2	Filter cloth washing	360
3	Boiler house and floor washing	230
4	Condenser water	1640

Characteristics

The characteristics of major fractions and of combined sewage excluding the condenser water are shown in Table below. Figures given in Table represent the average of several cane sugar industries in the country and those in Table give the total range of variation. It may be observed that there is wide variation in the values of the various parameters. As such actual survey of the conditions and analysis of sewage are necessary to determine the quantity and quality of the sewage to be handled in a particular factory.

Table: Characteristics of Effluents from Different Section of Cane Sugar Industry.

S. No.	Source of effluent	pH	Total solids (mg/l)	Suspended solids (mg/l)	BOD (5-day at 20° C) (mg/l)
1.	Mill house	6.7	1760	910	210
2.	Filter cloth washing	9.5	6970	4000	1765
3.	Boiling house	7.2	5130	120	5150

Table: Characteristics of Sewage (Excluding Condenser Water) from Cane Sugar Industry.

S. No.	Characteristics	Range
1.	pH value	4.6 to 7.1
2.	Solids (mg/L)	
	◦ Total	870 to 3500
	◦ Suspended	220 to 800
	◦ Volatile	400 to 2200
3.	Biochemical oxygen demand (5-day at 20°C), (mg/l)	300 to 2000
4.	Chemical oxygen demand (mg/L)	600 to 4380
5.	Total nitrogen (mg/L)	10 to 40

Pollutional Effects of Sewage from Cane Sugar Industry

The various pollutional effects of sewage from cane sugar industry are as indicated below:

- The effluents from some sections of the cane sugar industry contain considerable concentration of suspended solids, which deposit and cause blockage in drainage and ditches and also delayed pollutional effects because of slow decomposition of the settled matter.

- Effluent from cleaning operation and heat exchangers may be acidic, alkaline or neutral. These are damaging due to high salt concentration and toxic to aquatic life.

- The effluents have a high concentration of sugar and other carbohydrates. During anaerobic conditions, obnoxious odour develops in the contaminated stream. Such a septic condition results in production of hydrogen sulphide gas in contaminated water. The effluent is imparted black colour because of precipitation of iron by hydrogen sulphide as the stabilization proceeds.

- These effluents have a high biochemical oxygen demand. Discharge of these untreated effluents in water courses results in depletion of oxygen content, making environment unfit for fish and other aquatic life.

Methods of Treatment and Disposal of Sewage from Cane Sugar Industry

The various methods of treatment and disposal of sewage from cane sugar industry can be grouped under four main heads:

- Waste prevention at source,

- Disposal on land,

- Conventional biological treatment,

- Treatment in lagoons or stabilization ponds.

Waste Prevention at Source

The quantity of polluting material getting into sewage can be reduced by taking certain measures for proper operation and maintenance of the manufacturing plant. Prevention and treatment of sewage should begin within the industry itself. Good housekeeping, efficient operation of evaporators to reduce entrainment to the minimum and proper handling and storage of molasses to eliminate spillovers are important in this connection.

Routine inspection of different units, particularly pumps, conveyors, pipes and other vessels and judicious use of water in the mill reduces the problem arising from floor sweeping and washings. The filter cloth washings may be given short detention in a holding tank before being allowed to mix with effluents from other sections.

Disposal on Land

Use of sewage from cane sugar industry for irrigation is considered to be the best method of its disposal, after preliminary treatment to remove oil and suspended matter and correct pH value. However, the use of large fraction of this sewage for irrigation purposes may result in the development of serious odour problem.

Further unfiltered sewage may cause accumulation of sludge in the filed. This difficulty may be avoided by using only one third of available space each year for sewage disposal and using that portion for next two years for cultivation of grain or vegetable crops. In some cases depending on the soil texture and pH value, occasional furrowing of land may help to dissipate the odour problem to considerable extent.

Conventional Biological Treatment

The treatment of sewage from cane sugar industry by activated sludge process and trickling filter has been studied by many investigators, with varying degree of success. It may be necessary to add nutrients like nitrogen and phosphorus to bring it to the level of BOD : N : P to 100 : 5 : 1 in aerobic biological system.

However, it has been the general observation that biological treatment would require large and expensive plant as well as skilled supervision and would be difficult to operate since the sugar season lasts for a short period of only 4 to 5 months, and that also particularly during the winter.

Treatment in Lagoons or Oxidation Ponds

Treatment of sewage from cane sugar industry by lagooning or in oxidation pond has been found most feasible and economical. It has been found that in regions where winter conditions are not severe, the period of retention of sewage in lagoons or in oxidation pond is not too high and degree of purification is reasonable.

In this method sewage is treated in two stages, the first stage being primarily anaerobic digestion in open deep pond and the second stage being aerobic oxidation in open shallow pond. The first stage is usually preceded by equalization of flows and characteristics of the sewage in an equalizing cum digestion pond having a detention period of one day.

The anaerobic digestion pond may have a liquid depth of not less than 2.5 m and a detention period of 6 days. At an organic loading of BOD in the range of 0.24 to 0.32 kg/m^3/day in the anaerobic digestion pond, the reduction in BOD may reach the order of 60 per cent. The effluent from this pond is further purified in an aerobic oxidation pond having a liquid depth less than 1.2 m and a detention period varying from 10 to 12 days.

A BOD loading of 4750 kg/m^3/day may be applied in the oxidation pond. The BOD removal in the oxidation pond may exceed 70 per cent. The overall reduction in BOD in this treatment plant may be of the order of 90 per cent or more. The final effluent will have a BOD between 60 to 100 mg/l. However, even with this BOD the effluent cannot be discharged in inland surface waters and it may be necessary to dilute it to get the required value (30 mg/L). In case of land disposal for irrigation, the effluent need not be diluted.

Incorporation of a suitable algal culture tank for acclimatizing algae to the effluent and dosing algae continuously to the oxidation pond is useful in successful operation of the oxidation pond.

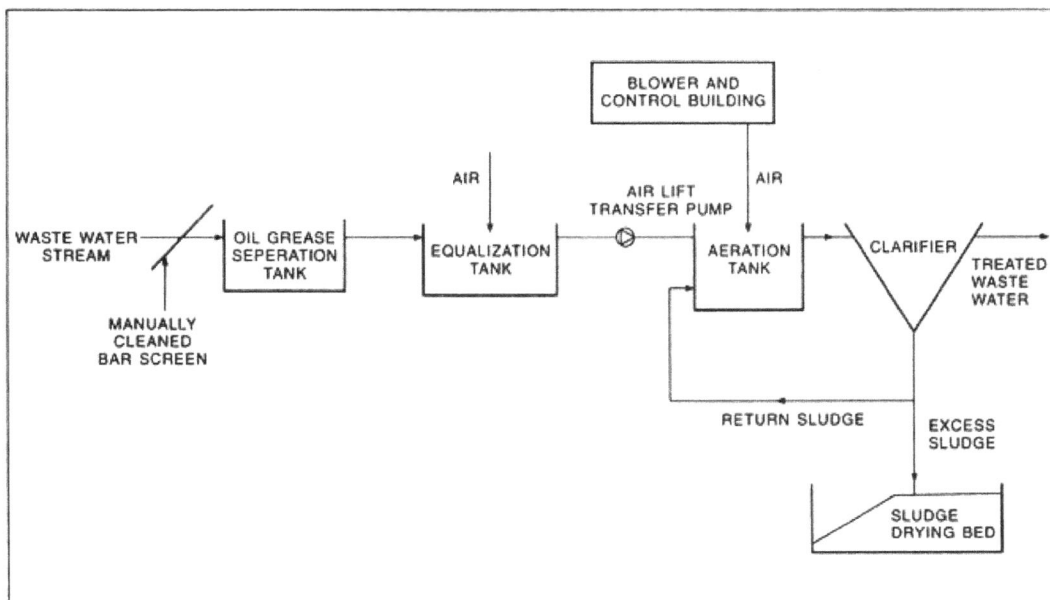

Flow diagram of effluent treatment plant in a sugar industry.

Byproducts of Cane Sugar Industry and their Disposal

The major byproducts of cane sugar industry are as follows:

- Bagasse,
- Press mud or Press cake,
- Molasses.

Bagasse

The production of bagasse depends upon the fibre content of the sugarcane, which varies from 13 to 15% in western and southern zone and from 15 to 17% in eastern and northern zone. Bagasse production is about 30% of the cane crushed.

The industrial use to which bagasse can be put are:

- Fuel,
- Raw material for manufacture of paper, pulp, newsprint and insulation board.

Bagasse can also be used after treatment for production of plastics moulding powder, cattle feed, biogas, manure, etc.

Press Mud or Press Cake

The production of press mud is about 3% of cane in sulphitation factories and 7% in carbonation factories. The sulphitation press mud also contains sugarcane wax varying from 8 to 10% of the mud.

The sulphitation press mud is almost entirely used as manure in the field, while carbonation press mud is generally used to fill up pits and in some cases for production of lime for building purposes. Attempts have been made to extract wax from sulphitation press mud, which can be used in place of imported carnauba required in manufacture of carbon paper, shoe and other polishes, wax paper, emulsion for protective coatings on fruits, etc.

Molasses

Molasses is the main byproduct of cane sugar industry. Production of molasses depends largely on the quantity and to some extent on the quality of the cane crushed which varies from region to region. Its average production is 4.4% of cane processed.

Molasses have an extremely high biochemical oxygen demand of the order of 900 000 mg/l, hence proper care should be taken so that there are no chances in the mill for the molasses to spill over into the sewage. Further considering the hazards of severe pollution of streams, special attention has to be paid to the disposal of molasses.

Molasses are mainly utilized by the distilleries. However, the production of molasses exceeds the amount required by all the distilleries of the country. About 70% of the molasses produced in the country is utilized by the distilleries.

Besides this, molasses is also used as:

- Cattle feed,

- Tobacco curing.

There is a good demand of molasses abroad and some quantity is exported.

References

- Industrial-wastewater-treatment: iwapublishing.com, Retrieved 15 April, 2019

- 7-common-types-of-industrial-wastewater-equipment: watertechonline.com, Retrieved 19 June, 2019

- Industrial-wastewater-reverse-osmosis-0001: wateronline.com, Retrieved 18 April, 2019

- How-to-manage-brine-disposal-and-treatment: saltworkstech.com, Retrieved 23 May, 2019

- How-to-treat-wastewater-from-fertiliser-industry-waste-management, water-pollution-wastewater-treatment-7488: environmentalpollution.in, Retrieved 08 January, 2019

- Wastewater-treatment-challenges-food-processing-agriculture: watertechonline.com, Retrieved 14 February, 2019

- How-to-treat-wastewater-from-cement-ceramic-industry-waste-management, water-pollution-wastewater-treatment – 7502: environmentalpollution.in, Retrieved 16 June, 2019

Chapter 6

Agricultural Wastewater Treatment

Agricultural wastewater treatment is the process of controlling pollution from surface run-off contaminated by chemicals present in fertilizers, pesticides and crop residues, etc. This chapter closely examines agricultural wastewater treatment to provide an extensive understanding of the subject.

Wastewater originating from agricultural processes is usually not contaminated with mineral elements, but instead primarily with significant loads of organic substances such as starches, sugar or cellulose.

Wastewater occurs at various stations on a farm. The areas of application include, above all, the cleaning of equipment and stables, as well as the cleaning of harvested products. A particularity is the treatment of silage leachate. To prevent contamination of the surface water by this highly polluted wastewater coming from silage production and storage, it must be separately collected and treated.

Flexible Wastewater Treatment for Agricultural Enterprises

Depending on the focus of the farms, differently polluted wastewater is produced in agriculture and must be treated before reintroduction into the cycle. If livestock are kept, large amounts of water accumulate when cleaning stables, milking equipment or milking parlors. Pig slurry, for example, can be fermented in suitable biogas plants. The resulting biomethane can be used to generate heat and power for the farm's own needs. The digestate is used instead of liquid manure as organic fertilizer on the farmland.

Silage leachate can also cause problems on some farms. It may also drip onto transport infrastructure during transport. These residues may mix with rain or melt water and significantly contaminate surface waters. Silage leachate should be contained and purified due to its high content of organic substances.

In the plant production sector, wastewater is produced primarily where products are cleaned and processed. As an example, today it is common to wash potatoes before bagging. Vegetables or fruit, such as apples, are also often washed before being packaged for sale. This results in wastewater, which requires treatment. Water is also essential for fruit harvesting and processing. As an example, up to five cubic metre of water are required to process one ton of potatoes. To reduce costs, water in many agricultural processing plants is increasingly recirculated in a loop. Impurities are expelled, and the water is reprocessed for reuse.

Silage Effluent Management

Silage effluent is a potent wastewater that can be produced when ensiling crops that have a high moisture content (MC). Silage effluent can cause fish-kills and eutrophication due to its high biochemical oxygen demand (BOD) and nutrient content, respectively. It has a high acidity (pH≈3.5-5) making it corrosive to steel and damaging to concrete, which makes handling, storage and disposal a challenge. Although being recognized as a concentrated wastewater, most research has focused on preventing its production. Despite noted imprecision in effluent production models-and therefore limited ability to predict when effluent will flow-there has been little research aimed at identifying effective reactive management options, such as containment and natural treatment systems. Increasing climate variability and intensifying livestock agriculture are issues that will place a greater importance on developing comprehensive, multi-layered management strategies that include both preventative and reactive measures.

One common side effect of silage making is silage effluent. No matter our best efforts to harvest at the correct moisture and in a timely fashion, it happens. If you have a silo, upright or horizontal, you have some amount of silage effluent.

Where that effluent ultimately flows to is the issue. From a pollution potential standpoint, silage effluent ranks among the highest sources. A common measure of pollution potential is the biological oxygen demand (BOD). Simply stated this is the amount of oxygen needed to break down the product. Silage effluent often has a BOD around 50,000 mg of oxygen per liter of effluent, while raw domestic sewage runs about 500 mg/L. The actual numbers are not important here, the point is that silage effluent is 100 times as strong as raw sewage. For a better example, as little as one gallon of silage effluent can lower the oxygen content of 10,000 gallons of fresh water to a critical level with respect to fish survival. The nutrient concentration of silage effluent, in terms of N, P and K, is very similar to typical liquid dairy manure. The pH is typically in the 4.0 range which is another potential pollution problem and leads to the characteristic vegetation burns around silos.

So what can be done in terms of treatment/disposal? First, think about placement and preparation of the silo. Silos should be located away from open waterways and wells. An unused piece of pasture down by the stream may seem like a nice place for a bunker, right up until the stream becomes polluted. Proper preparation of the site will also help with collection and/or treatment of the effluent. Divert clean water away from the site and layout the site so on-site runoff will move toward a common point for collection and/or treatment.

To limit the treatment, try to minimize the amount produced. Basically this means ensiling at the proper moisture content. When stored at a dry matter content of 30% effluent flow is greatly reduced. Covering horizontal silos also helps reduce the prolonged flow of effluent.

But, even when properly located and minimized, there is still effluent to handle. There are several methods to handle effluent. Probably the most often used method is incorporation with a liquid manure system. Be very careful when mixing silage effluents with manure. Hydrogen sulfide and other poisonous gases are produced, and it should not be done if the storage is covered or under the barn. The second option is land application to a crop field or grass filter strip. However, effluent must first be diluted with water if applied to a growing crop; typically a 1-to-1 ratio will work. The effluent can also be directly applied to fields with a non-growing crop.

Treatment must be addressed on a case by case basis, because each farm is different.

Animal Farm Wastewater Treatment System

Animal farm wastewater mainly comes from livestock and poultry manure and washing wastewater, which contains a large number of pollutants, such as heavy metals, residues of veterinary drugs, and a lot of pathogens, etc. In addition, livestock and poultry breeding wastewater contains much nitrogen, phosphorus, potassium and other nutrients.

Animal farm wastewater belongs to high concentration organic wastewater which is rich in a large number of pathogens (escherichia coli, hatched from eggs, etc.). It is inappropriate to directly discharge into water or storage place. The rain flushing into the water, it may cause serious deterioration of surface water and groundwater quality. Due to the strong leaching of livestock and poultry manure, and large amounts of nitrogen, phosphorus and water soluble organic matter in manure leaching, if not properly handled, it enters through the surface runoff and infiltration into the underground water, causing the pollution of groundwater.

Main influence for surface water is, that a large number of organic matter enters into water, after the decomposition of organic matter it will deplete the dissolved oxygen in the water, making water stink; if human and animal drink for a long time, it will cause the poisoning. And some of the toxic algae growth and large amounts of multiplying can discharge toxins in the water, causing massive death of aquatic animals, which seriously damages the water ecological balance; some of urine bacteria and viruses can lead to the spreading of some epidemic with water flowing etc.

Animal Farm Wastewater Treatment Process

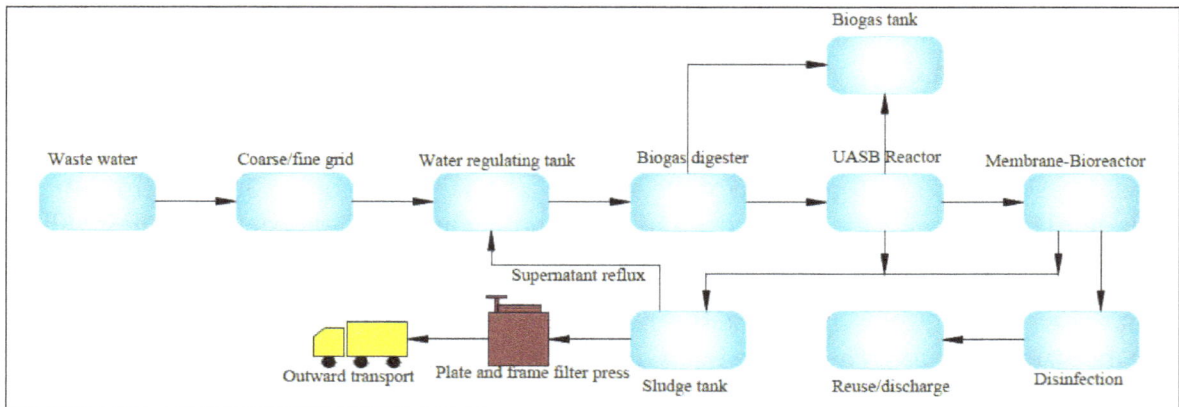

Water quality characteristics of animal farm wastewater is that it contains various kinds of organic substances, feces and sandy soil pollutants. This kind of wastewater treatment can achieve good treatment effect by combining biochemical and physicochemical.

Animal farm wastewater out from pig farms enters into the tank after flowing through the grid, the mesh to remove most large size of sundry. In the biogas tank, organics conduct hydrolysis under the conditions of anaerobic, bio-refractory macromolecule organics being transformed into bio-degradation micro-molecule organics for the follow-up process. Biogas tank produces a certain amount of biogas, collected by biogas equipment for recycling. Animal farm wastewater is sent by the lift pump into the hydrolysis acidification tank after balancing water quality and water yield in the adjusting tank. Hydrolysis acidification tank adopts up-flow anaerobic sludge bed (UASB) reactor, the sewage from bottom up across the sludge bed formed by the microorganism in the tank, (organic) pollutants in wastewater is trapped by the sludge bed, macromolecule and complex organics are resolved into low simple molecular organics (such as organic acid) by decomposition, absorption and decomposition. At the same time, the mixture contacting oxidation tank back conducts denitrification under the action of denitrifying bacteria to achieve the action of denitrifying bacteria.

Wastewater out from UASB reactor enters into the MBR. MBR is composed of oxygen and MBR reaction tank. Water production in the MBR tank can meet the standard for discharge or reused in piggery back flushing after sterilization killing the bacteria in the wastewater.

Treatment of Pesticide Containing Wastewaters

The methods for the disposal of low level pesticides include: land cultivation, disposal pits, evaporation ponds and landfills. There are three types of disposal pits: soil pit, plastic pit and concrete pit.

Land Cultivation

In this method, excavated contaminated soil is spread out in a thin layer on uncontaminated soil in order to allow for natural chemical and biological processes to transform and degrade the contaminants. Soil contains microbes (fungi, algae and bacteria) capable of metabolizing pesticides. The ability of bacteria to metabolize pesticides has been well documented by several researchers. Bhadhade et al. reported that soil bacteria was capable of degrading 83% - 93% of the organo-phosphorous pesticide monocrotophos. Ohshiro et al. reported a 96% reduction in isoxathion from the organophosphouruspestiside by bacteria isolated from turf green soil. Kearney et al. reported that soil microbes were capable of degrading 90% of the alachlor pesticide within 30 - 40 days. Tang and You reported that the triazophos bacteria was capable of degrading 33.1% - 95.8% of pesticides in soil.

Table: Disposal methods of pesticide containing wastewater.

Method	Description	Advantages	Disadvantages
Land Cultivation	Place liquid wastes in plow zone of soil for subsequent weathering.	On-site use Simple technology.	Land requirements Possible runoff and leaching Slow and variable decomposition Restricted vegetation.

Disposal Pits	Place liquid wastes in pits containing soil and open to air for subsequent weathering.	On-site use Simple technology Secure containment.	Slow decomposition Limited lifetime of pit Effectiveness varies with climate.
Evaporation Ponds	Place liquid wastes in lined ponds open to air for subsequent weathering	On-site use Simple technology Secure containment	Slow decomposition Limited lifetime of pond effectiveness Varies with climate.
Landfills	Burial of wastes in soil.	Generally available Complete removal.	Land requirements High transportation costs Possible runoff and leaching.

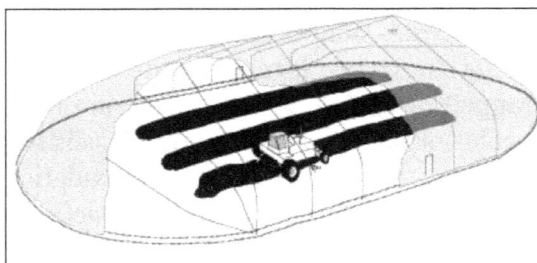

Land cultivation of contaminated soil.

Racke and Coats reported that after soil has been treated with a pesticide a few times its microorganisms build up a need for that pesticide which results in fairly rapid degradation of any additional applications. Schoen and Winterlin stated that natural soil degradation is effective when low concentrations of the pesticide are present, but with high concentrations of pesticide it becomes much more difficult to degrade. Felsot stated that land cultivation is only effective for compounds that can be biotransofrmed or biominerlized by soil microbes. Somasundaram et al. reported that the ability of soil microbes to degrade certain pesticides is affected by pesticide toxicity to soil microbes that are responsible for the degradation. Felsot et al. stated that land cultivation is effective if the pesticide is degraded at the same or faster rate than it is applied to the field. Felsot et al. noted that land cultivation can be enhanced by the addition of organic amendments such as sewage sludge.

Soil Pit

A primary method for disposing of liquid pesticide waste is by dumping it in an unlined soil evaporation pit, usually 15 × 15 × 1 m Schoen and Winterlin reported that factors such as chemical structure and concentration of pesticide play a major role in the degradation of pesticides in soil pits. Gan and Koskinen stated that the dissipation of the pesticide decreases as the concentrations of pesticide increases. Dzantor and Felsot and Gan et al. noted that high pesticide concentrations may cause microbial toxicity which would inhibit the degradation of the pesticide.

Several researchers noted that the prolonged dissipation of pesticides opens a window for runoff and leaching, especially at higher pesticide concentrations. Gan et al. reported that 50% dissipation of the alachlor pesticide in soil, at concentrations of 4 and 4 300 mg/kg took approximately 2 and 52 weeks, respectively. Gan et al. noted that atrazine pesticide took approximately 4 and 24 weeks to be dissipated to half the concentration of 7 and 6400 mg/kg in soil, respectively. Schoen and Winterlin noted that captan, trifluralin and diazinon at concentrations of 100 mg/kg took 1 - 2, 116 - 189 and 77 - 160 weeks to dissipate to half the concentration while captan, trifluralin and diazinon at concentrations of 1000 mg/kg took 30 - 48, 168 - 544 and 77 - 160 weeks to reach 50% disappearance in soil, respectively.

Plastic Lined Pit

This method for disposal of pesticide waste requires proper selection of the site to avoid leaching and runoff.

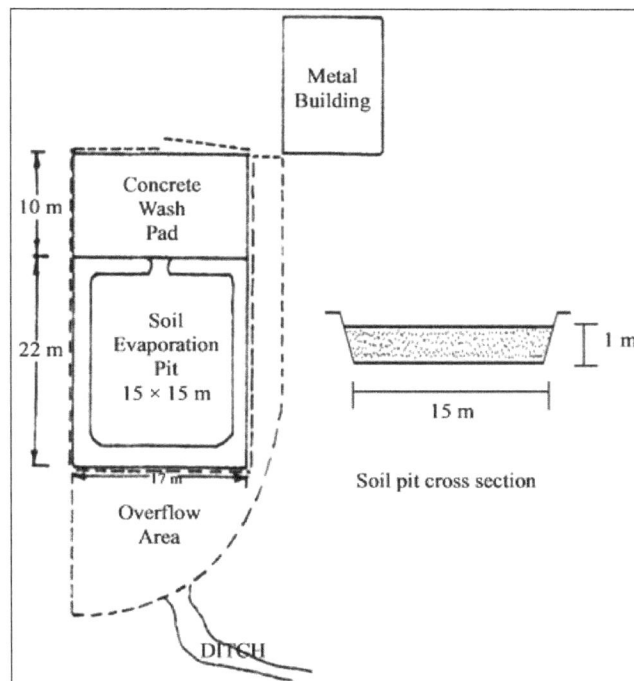

A soil pit for disposal of pesticide water.

The site should be in an area where there is no danger of contaminating dwellings groundwater sources and surface water used for crop and livestock production. The pit should be on a levelled ground with a depth of 0.5 - 1 m covered with a plastic liner and a layer of soil is laid on top of the liner. The pit should be open to the atmosphere in order to allow for water evaporation into the atmosphere. A roof cover will prevent the water level from raising due to rain or snow. The wastewater is pumped into the pit for pesticide biodegradation by soil microbes Hall et al. reported that the presence of microbes in the soil water mixture in plastic lined pits was responsible for the degradation of pesticide and no accumulation of pesticide was noted in the pits. Junk and Richard evaluated the effectiveness of 90,000 L polyethylene lined disposal pit with over 150 kg of 25 different types of pesticides for over 2 years and concluded that this method was in fact effective for disposal of pesticide waste with insignificant release to air and water surroundings.

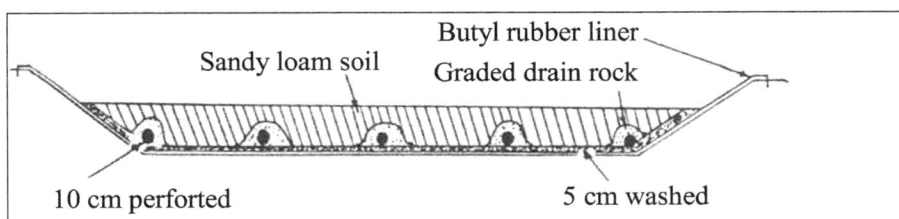

A plastic lined evaporation pit for disposal of pesticide wastewater.

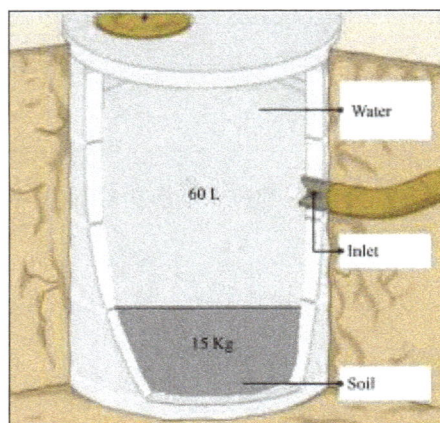

A cross section of a plastic pit for disposal of pesticide wastewater.

Figure illustrates a cross section view of a simple small scale plastic pit used to dispose of pesticide waste. It consists of a plastic drum with a length and width of 75 × 55 cm, respectively. Inside the drum is a mixture of 15 kg of soil and 60 L of water, that was used to treat pesticide waste which was introduced into the system through the inlet. Junk et al. used 56 plastic containers filled with 15 kg of soil and 60 L of water to test the degradation of alachlor, atrazine, triflualin, 2,4-D ester, carbaryl and parathion and found this system not suitable for atrazine but was effective and very rapid for 2,4-D and carbayl. They concluded that: 1) the plastic container provided satisfactory containment for most common pesticides, 2) soil was a satisfactory source for microorganisms, 3) aeration and buffers had questionable value, 4) half life concept for degradation was not applicable and 5) sampling from small disposal sites was a problem.

Concrete Pit

Similar to the plastic lined pit, the concrete pit should be on levelled ground with a depth of 0.5 to 1 m, a length of 8 - 10 m and a width of 3.5 m and reinforced with 0.20 m thick concrete walls. The pit consists of a top and bottom layer of gravel that is 4 cm in diameter with the middle layer consisting of topsoil. The pit should also have a cover to prevent rise in water level from rain or snow but remain open to the atmosphere in order to allow for water evaporation.

Johnson and Hartman tested the microbiological activity in a concrete pit and concluded that the degradation process in the pit was effective and no long-term accumulation of pesticide was present. Junk and Richard tested the effectiveness of 30,000 L concrete disposal pit with over 50 kg of 40 different types of pesticides for 8 years and concluded that this method was in fact effective for disposal of pesticide waste with insignificant release to air and water surroundings. Hall tested the effectiveness of an open concrete disposal pit for the degradation of 45 pesticides over five months and concluded that the biodegradation of the pesticides was successfully accomplished and the pit did not leak or pollute the air, but the system was too large and complicated for most farms.

Evaporation Beds

Lined evaporation beds are used for the disposal of pesticide wastewater. Leach lines underneath the soil surface supply the beds with the pesticide residues from washing equipment. The pesticides rise to the beds surface where they are degraded through photochemical, chemical and biological actions and are distributed via air vapour. Some of the beds have hydrated lime incurpo-rated into the soil in order to aid in the degradation of certain pesticides. A medium scale disposal system of this type costs up to $50,000 to construct.

Hodapp and Winterlin reported a reduction in the diazinon pesticide of 62.54% using lined evap-oration bed without lime and a degradation of 77.75% with lime. They also reported an ethyl para-thion reduction of 69.83% using lime treatment in the beds and a reduction of 45.45% without the use of lime. Winterlin et al. tested ten (6 × 12 × 1 m) lined (with a butyl rubber membrane liner and 36 cm of sandy loam soil) evaporation beds to determine their pesticide decay effectiveness. Pesticide rainsate was introduced through subsurface tiles in limited amounts and the effects of geography, climate and lime application were examined. The method appeared to be beneficial for disposal of some pesticides but not all. Over 100 pesticides were tested, but only 46 were actually detected.

This method for the disposal of pesticide containing wastewaters is advantageous because the beds are economical, little maintenance is required, do not build up high levels of pesticides and are effective in degrading as well as containing the pesticides without excessive exposure through air vapour. It is considered an economical, on-site method of disposal which requires only annual monitoring. The disadvantages appear to be the development of a high concentration of residue in the top layer of the soil and the difficulty in acquiring a representative sample.

Land Filling

Landfills are sites that dispose of waste by burial into the soil where microorganisms are used to change the composition of the toxic elements. Landfills for pesticides are equipped with drying pits containing soil to provide the microbes needed to break down the pesticide components into non harmful elements. A nearby sump for the propose of draining and rinsing the containers that have not been fully emptied or rinsed.

Munnecke reported that soil bacteria were capable of hydrolyzing ethyl parathion found in pesti-cide container residues within 16 h. Johnson and Lavy reported that carbofuran, thiobencarb and triclopyr buried in degrading containers dissipated to 50% of the initial concentration with the first 94 days or less, while benomyl took 179 - 1020 d before 50% dissipation. They also noted that the rates of dissipation decreased with an increase in soil depth.

Yasuhara et al. detected 190 compounds in landfill leachates in Japan. Williams et al. reported that the pesticide mecoprop is found in landfill leachate because it is resistant to anaerobic degrada-tion. Christensen et al. noted the presence of the pesticide bentazon, N,N-Diethyltoluamide and mecoprop in landfill leachate because of their persistence to anaerobic landfill conditions. Alloway and Ayres noted the presence of the pesticides atrazine and simazine in landfill leachate.

Pesticide Treatment Methods

The pesticides treatment methods include: 1) thermal treatment, 2) chemical treatments, 3) phys-ical treatments and 4) biological treatments. Thermal treatments include incineration and open burning. Chemical treatments include ozonation/UV radiation, Fentonoxidation, hydrolysis and KPEG. Physical treatments are based on absorption using activated carbon, inorganic and organic materials. Biological treatments include composting, phytoremediation and bioaugmentation.

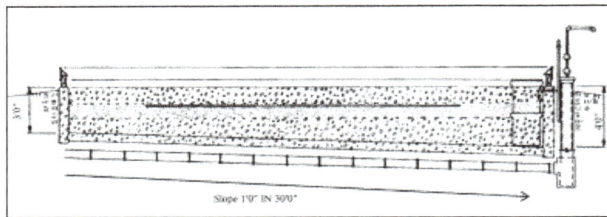
A cross section of concrete pit for disposal of pesticide wastewater.

Evaporation beds for disposal of pesticide wastewater.

A landfill for disposal of pesticide wastes.

The landfill leachate collection pit.

Incineration

Pesticide Incineration is a high temperature oxidation process where the pesticide is converted into inorganic gases (water vapour, CO_2, volatile acids, particles and metal oxides) and ash. Incineration of pesticide should be operated at temperatures higher than 1000°C so that the pesticide can be treated within the first 2 seconds. At such temperatures, smoke production is nil and the generated combustion gases are similar to those generated by wood burning. Temperatures lower than 1000°C can also be used as long as the incineration time of the pesticide does not exceed 2 seconds. However, lower temperatures tend to produce toxic intermediate products.

Kennedy et al. noted change in the combustion efficiency over the temperature range of 600°C - 1000°C. Ferguson and Wilkinson reported that incineration has 99.99% destruction efficiency at temperatures of 1000°C and a retention time of 2 s in the combustion zone. Steverson reported a 99% destruction efficiency for 16 currently used insecticides and herbicides at temperatures ranging from 200°C to 700°C. Linak et al. reported an incineration efficiency of greater than 99.99% for dinoseb. Ahling and Wiberger noted that the incineration of fenitrothion and malathion at temperatures lower than 600°C gave emissions of 1% - 2% of the pesticide amount added and temperatures above 700°C would be required to achieve safe destruction and emission levels.

Table: Current treatment methods of pesticide containing wastewater.

Method	Description	Advantages	Disdvantages
Thermal	Controlled combustion of either liquid waste or concentrated residue.	Destructive Rapid No by-products.	High costs Complex Not useful for some chemical.
Chemical	Chemical destruction through use of oxidative, reductive, hydrolytic or catalytic reagents.	Destructive Rapid.	High costs Complex Variable effectiveness.

Physical	Removal of chemicals from waste-water by adsorption and/settling.	Rapid Possible on-site use.	No destruction involved By-products for disposal.
Biological	Use of micro-organisms to destroy chemicals.	Destructive.	High costs Susceptible to shock Relatively slow Variable effectiveness.

Incinerators capable of achieving high levels of destruction are equipped with a combustion chamber, an afterburner, scrubbers and electrostatic filters. Ferguson and Wilkinson reported the following performance standards for incinerating hazardous wastes: 1) the incinerator must achieve a destruction and removal efficiency greater than 99.99% for each of the chemicals present in the waste feed, 2) HCl emissions must not exceed 1.8 kg/h or 1% of the HCl in the stack gas prior to entering any pollution control equipment and 3) the particulate matter emitted must not exceed 180 mg/DSCM when corrected to 7.0% O_2. The advantages of incineration include: 1) effectiveness in degrading chlorinated organics, 2) destruction efficiency of 99.99% and 3) setup at locations next to plants generating the waste. The disadvantages of incineration technology include: 1) need for sophisticated equipment 2) production of cyanide in the off gas during the incineration of organonitrogen pesticides, 3) too costly and complex, 4) it is intended for centralized large scale disposal and 5) not recommended for inorganic pesticides.

Open Burning

This method combusts pesticides and pesticide waste containers by piling up empty paper and plastic containers and setting them on fire. Although this method is inexpensive and convenient, it is hazardous to workers, plants and animals. It is prohibited in some cases by the Regional Air Quality regulations in the US. It emits gases, smoke and fumes into the atmosphere as well as toxic residues that are left in the containers.

Adebona et al. noted several products of incomeplete combustion, polyaromatic hydrocarbons and low levels of dioxins in open burning tests on 22.7 kg insecticide bags. Oberacker et al. noted that after burning bags containing phorate, 2% of the phorate was released into the air and 0.5% remained in the solid residues. Felsot et al. reported that bags containing atrazine released 13% of the remaining product into the air while 25% remained as residue. Such results indicate that the temperatures for complete combustion were not reached or were not maintained long enough in order to obtain destruction efficiencies of 99.99% or greater.

Ozonation/UV Radiation

The use of ozone and UV radiation to enhance the oxidation of aromatic compounds was investigated by several researchers. Ozonation is more effective treatment method in the presence of UV light because it canform hydrogen radicals which are very effective oxidizing agents. The benefits of this process are its mobility, ease of operation and rapid effects. The disadvantages are its high energy consumption and initial equipment cost.

Kuo used a UV/ozonation system consisting of a medium-pressure mercury vapor lamp with a water cooling jacket and an ozone generator. The lamp power consumption was 150 W and was capable of 14.3 W output (at 3.0 mW/cm2 at a distance of 9 cm). The O3 was pumped at a rate of 400 mg O3/hr/L solution. A solution of 2% KI was used for absorbing the residual ozone from the reactor.

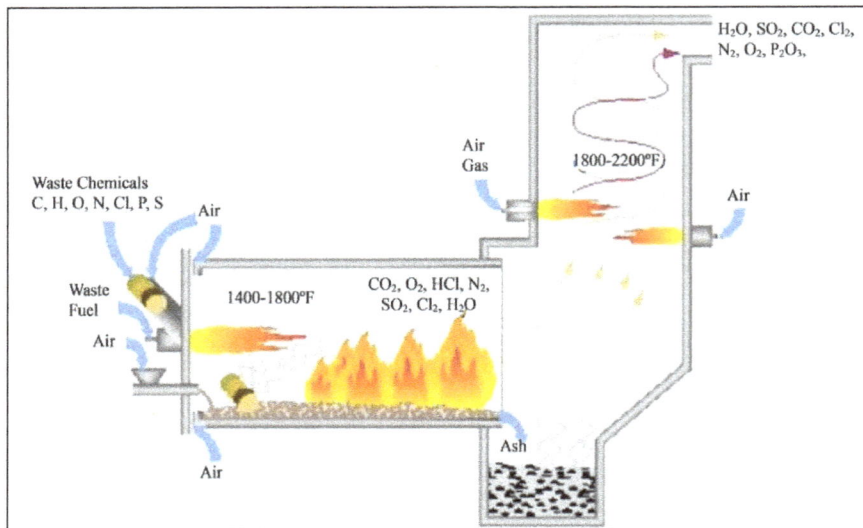

An incinerator for pesticide wastes.

Open burning of pesticide wastes.

Somlich et al. noted that irradiation of the alachlor pesticide achieved de-chlorination of the compound, while ozonation works to oxidize the compound into several intermediate products. Under Ultraviolet irradiation, the photon absorption by the carbonyl present in the compound is then followed by the loss of the chlorine. The pesticide degradation reactions that take place under UV/ozonation are as follows:

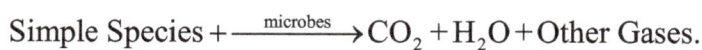

$$\text{Pesticide} + O_3 \xrightarrow[\text{H}_2\text{O}]{\text{UV light}} CO_2 + H_2O + \text{Simple Species},$$

$$\text{Simple Species} + \xrightarrow{\text{microbes}} CO_2 + H_2O + \text{Other Gases}.$$

Kearney et al. monitored the degradation of alachlor using a UV/O_3 system by measuring the concentration of the 14 CO_2 released. Under UV radiation, the alachlor pesticide was completely depleted from the water with the presence of oxygen within 25 minutes while it and took 50 minutes before it was fully depleted with ozone alone.

Fenton Oxidation

The Fenton process can be used as part of an oxidative system to treat and degrade pesticides. It consists of hydrogen peroxide (H2O2) and iron salts at low pHs. The iron salts act as a catalyst, increasing the effectiveness of the H2O2 by forming highly reductive hydroxyl radicals. The radicals are capable of oxidizing other species that are present in the solution as follows.

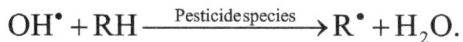

$$H_2O_2 + Fe^{2+} \rightarrow Fe^{3+} \, OH^- + OH^{\bullet},$$

$$OH^{\bullet} + RH \xrightarrow{\text{Pesticide species}} R^{\bullet} + H_2O.$$

Hydroxyl radicals are very powerful oxidizing agents with a 2.33 V oxidative potential. The rate of degradation of organic pollutants is strongly accelerated by UV irradiation. The photolysis of the Fe3+ complexes allows the regeneration of Fe2+ thus allowing the reaction to proceed much quicker in the presence of H2O2. The advantages of this method for pesticide treatment are: low cost, ease of operation, simplicity and the wide range of temperature that can be used.

Fallmann et al. noted a 72% reduction in 100 ppm total organic carbon solution using 23 mL of hydrogen peroxide and a reaction time of 124 minutes in a photo assisted Fenton process. Larson et al. reported that in the presence of ferric perchlorate and a mercury lamp, the atrazine pesticide had a half-life of less than 2 minutes compared to 1500 minutes when iron salt was not present. Huston and Pignatello noted a half-life of less than 10 minutes for the captanpesticide using UV assisted Fenton reagent at a pH of 2.8. Pignatello and Sun reported a half-life of 2 minutes for methyl parathion using UV assisted Fenton reagent. Doong and Chang reported a half-life of less than 10 minutes for alachlor pesticide under photo assisted Fenton reagent at a pH of 2.8.

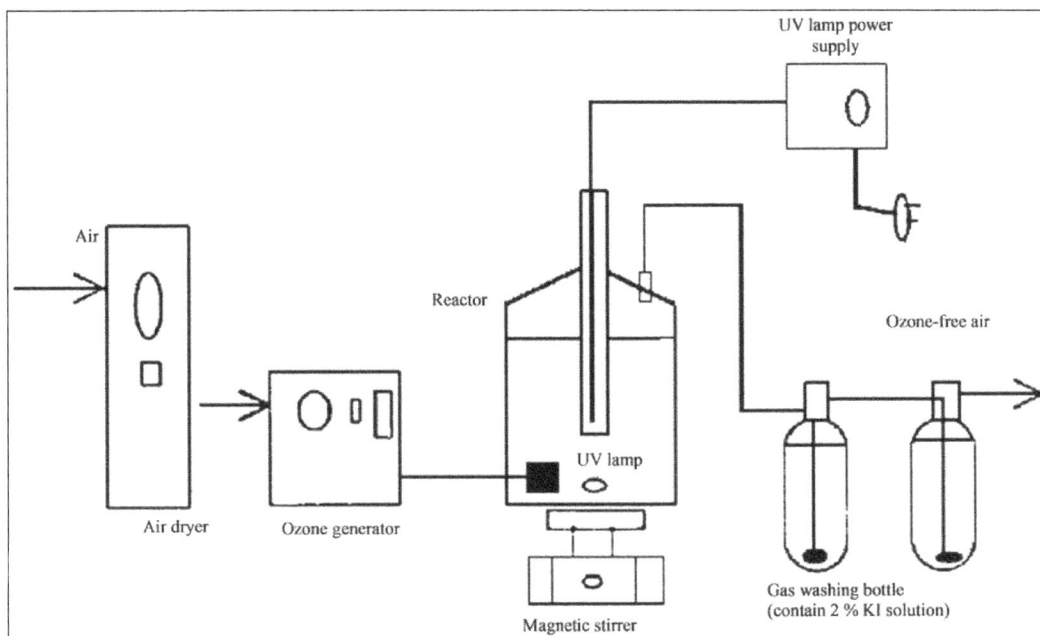

A UV/O3 system for the treatment of pesticide wastewater.

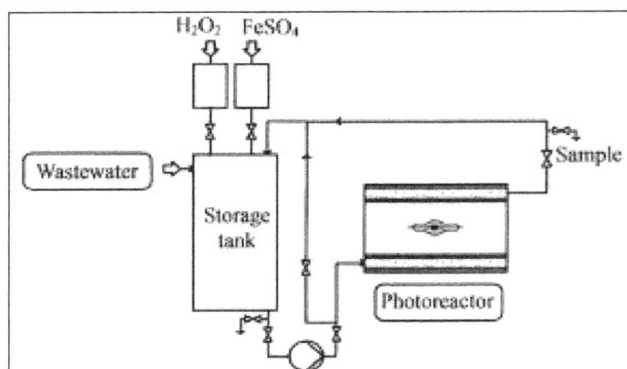

A photo assisted Fenton system for treatment of pesticide wastewater.

Hydrolysis

This method for pesticide treatment works by hydrolyzing the ester linkages found in pesticide compounds, including pyrethroids, carbamates, organophosphates and acetaniledes. These compounds can be hydrolyzed in solutions with high pH levels. Desmarchelier used calcium hydroxide for ester hydrolysis and found it to be a safer alternative to sodium and potassium hyoxides for the hydrolysis of fenitrothion pesticide. Lee et al. noted that under basic conditions, sodium perborate was more effective in the hydrolysis of organophosorus than sodium hydroxide, because the peroxide anion released from sodium perborate is much more reactive to organophosphorus than the hydroxyl ion. Qian et al. noted an enhancement in the hydrolysis process of mevinphos, diazinon, methyl parathion, malathion and parathion in lake water (10 mg/L) using sodium perborate at pH of 9.88. However, with the presence of soil, the reaction was noted to be significantly slower and the concentration of perborate had to be increased by four folds.

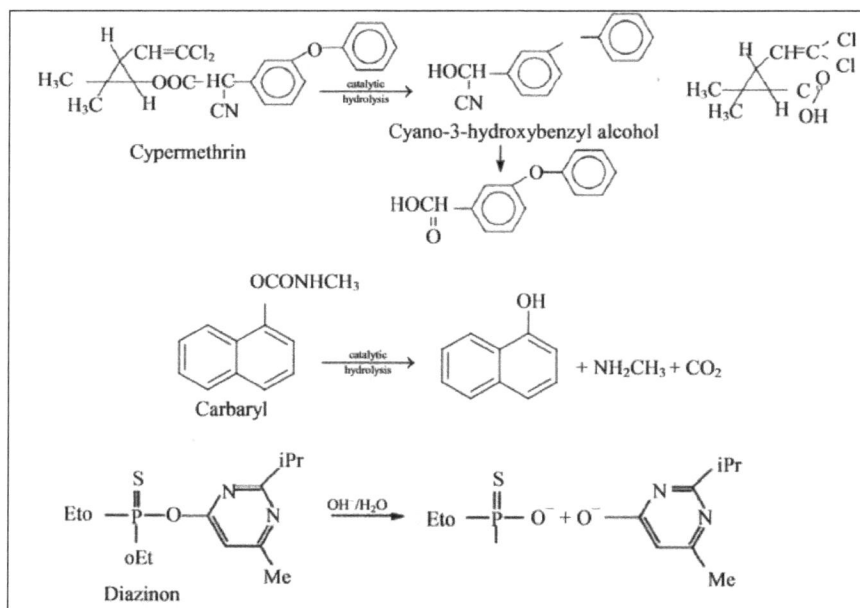

Hydrolysis of cypermethrin, carbaryl and diazinon pesticides.

Metal oxide and divalent metal ions have been noted for their ability to catalyze the hydrolysis of organphosorus insecticides. Smolen and Stone reported that the phophorothionate insecticides (chlorpyrifos-methyl, zinophos, diazinon, parathion-methyl and runnel) and phosphorooxonates (chlorpyrifos-methyl oxon and paraoxon) were most effectively catalyzed by Copper (II). The downside of catalysis using metal oxides is the formation of products with significant toxicity. Badawi and Ahmed noted that the hydrolysis of the pesticides diazinon, cypermethrin and carbaryl was effective and accelerated by the addition of a copper (II) ion complex.

KPEG

Potassium polyethylene glycol ether (KPEG) is capable of destroying chlorinated pesticides. Chlorinated hydrocarbons and cyclodienes are resistant to degradation by hydrolysis. Dechlorinating these pesticides with KPEG would then enable their biodegradation through land treatment processes. KPEG was found to be capable in dechlorinating polychlorinated biphenyls (PCBs) in soil and solvents. In older formulations of phenoxy herbicide, KPEG was found to be capable of degrading dioxins and dibenzofurans. The reaction that takes place consists of a nucleophilic substitution and a phase transfer at the carbon-halogen bond as illustrated by the following equations.

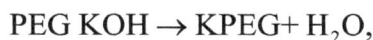

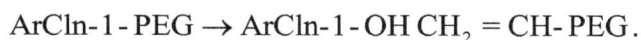

$$PEG\ KOH \rightarrow KPEG + H_2O,$$

$$KPEG\ ArCln \rightarrow ArCln\text{-}1\text{-}PEG\ KCI,$$

$$ArCln\text{-}1\text{-}PEG \rightarrow ArCln\text{-}1\text{-}OH\ CH_2 = CH\text{-}PEG.$$

where:

PEG = polyethylene glycol monomethyl ether

Ar = aromatic nucleus

Taylor et al. reported that the vessel for the KPEG reaction consists of a 55-gal drum (surrounded with heat tape capable of maintaining the temperature at 70°C - 85°C) and an electric motor with a mixer. With the reagents KOH and PEG, this vessel was capable of degrading 98% of phenoxy herbicide waste. The generated waste contained in the drums can remain there, eliminating the need for transfer into another container. Vapor emitted from the reaction drums are condensed in a water drum, the remaining condensables are traped in the scrubber containing sodium hypochlorite solution. Vapors are then passed through an activated carbon absorbent and as well as a molecular sieve.

The materials and chemicals needed for the KPEG process are easy to find. The disadvantage of the KPEG process are 1) high clay content, acidity and high natural organic matter interferes with KPEG reaction and 2) its not recommended for large waste volumes with concentrations above 5% for chlorinated contaminants. If necessary, emissions can be controlled by construction of a vent system with scrubber and absorbent.

Inorganic Absorbents

Pesticide adsorption can be performed using anionic clays (layered double hydroxides) which are simple to prepare, hydrotalcite, which occurs in nature may also be used as a layered double hydroxide (LDH). A variety of compounds can be formed by changing the cation metal. In order for a material to be considered as a good adsorbent it must possess the following properties: 1) a granular structure, 2) insoluble in water, 3) chemical stability and 4) have a high mechanical strength. Figure is multi-functional gravity filter which can be used for various water treatment methods by employing various adsorbent media.

Niwas et al. reported that styrene supported zirconium (IV) tungstoophosphates was successful in adsorbing the pesticide phosphamidan. Inacio et al. noted that the inorganic adsorbent Mg3AlCl was capable of adsorbing the MCPA herbicide within 30 - 45 minutes at room temperature. Boussahel et al. noted a removal efficiency in ayanaz in and atrazine of 85% - 90% using $CaCl_2$ or $CaSO_4$. Bojemueller et al. reported that the pesticide metolachlor can be adsorbed by bentonites and the adsorption efficiency can be doubled by increasing the temperature. Li et al. noted that the pesticide glyphosate was adsorbed on the external surface of MgAl-LDH at low concentrations, while at high glyphosate concentrations an inter layer ion exchange occurred.

Organic Absorbents

Various organic materials can be used as good adsorbents for pesticide removal. Ahmaruzzaman and Gupta reported that rice husk is insoluble in water, possesses an irregular granular structure and has a high mechanical strength and chemical stability that make it a good adsorbent. Chowdhury et al. noted that treated rice husk was capable of removing 89% - 97% of malachite green pesticide. Akhtar et al. investigated the adsorption potential of selected agricultural waste materials (rice, barn, bagasse fly ash from sugarcane and rice husk) for the pesticide removal of methyl parathion from wastewater and reported pesticide removal efficiencies in the range of 70% - 90% within 90 minutes. Memon et al. reported that thermally treated watermelon peels were capable of removing 99% of the methyl parathion pesticide. Al hattab and Ghaly reported a captan removal efficiency of 99.2% and 98.5% using hay and soybean plant residues, respectively.

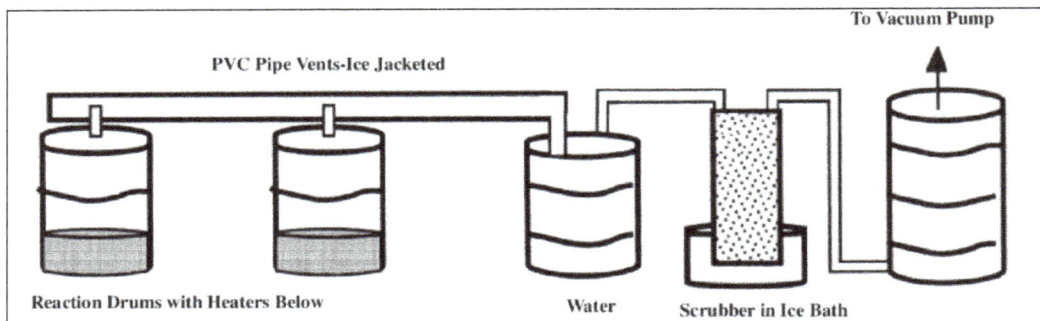

A KPEG process with vent system.

Filters used to treat pesticide wastewater through adsorption.

Activated Carbon

Carbon adsorption treatment method for pesticide containing wastewater is used in the pesticide manufacturing industry as well as in pesticide cleanup. The activated carbon system consists of a prefilter made up of sand or an alum flocculation chamber with a carbon filter. Dennis and Kobylinski reported on a Carbolator system which uses a suspended bed of carbon packed in bags of floating porous polyethylene in order to avoid clogging. The water was continuously recirculated through the carbon filters by directing it back into the waste holding tank.

Felsot et al. reported that rinsewater containing malathion, propoxur, chlorpyifos, diaxinon and dimethoate were all removed to nondetectable levels using the Carbolator. This process reduced the amount of waste generated by several magnitudes through efficiently absorbing pesticides form the water. Kobylinski et al. used a Carbulator 35B to remove baygon, dimethoate, diazinon, runnel, malathion, dursban and 2,4-D and found that the higher the molecular weight of the compound the more favourable the effect of adsorption by activated carbon. Similar findings were also reported by other researchers.

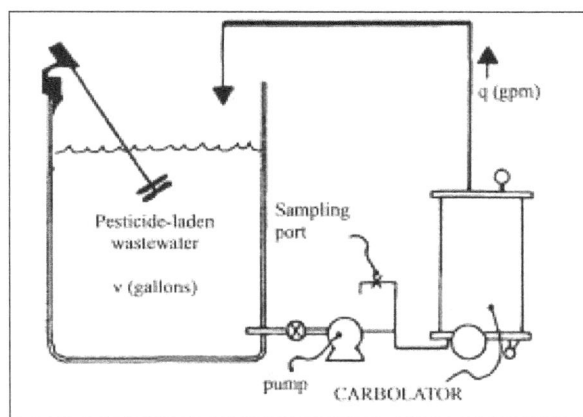

Recirculation through activated carbon.

Honeycutt et al. reported that a waste stream containing 100 ppm chlorophenols was reduced to 1 ppm using activated carbon. Giusti et al. reported a carbon activated adsorption of 3.6% and 98.5% for methanol (molecular weight of 32 g/mol) and 2-ethyl hexanol (molecular weight of 130.2 g/mol), respectively. Sarkar et al. reported an adsorbent efficiency of 98% - 99% for the removal of the isoproturon pesticide using powdered activated charcoal. Gupta et al. reported an adsorption efficiency of 70% - 80% using activated charcoal for removing pesticides. Word and Getzen reported that a decrease in pH increased the adsorption of aromatic acid compounds due to enhancement of carbon surface properties.

The activated carbon is very effective in removing pesticides and it does not require extensive monitoring. The disadvantages include: 1) the need for a skilled chemist for field testing, 2) the high cost and 3) this process is only capable of adsorbing solutions with concentrations of less than 1000 ppm.

Composting

This treatment method relies primarily on microbial activity and aeration efficiency. Microorganisms that are naturally occurring in the materials increase significantly in numbers and begin to decompose biodegradable compounds which results in the release of carbon dioxide as well as the production of metabolic heat, causing the temperature of the compost to rise to 60°C - 70°C. As the compost temperature increases, three succession of microbes occur: psycrophilis, mesophilis and thermophilis.

Racke and Frink reported a complete degradation of carbaryl during the composting of sewage sludge. Petruska et al. achieved a complete degradation of diazinon pesticide using dairy manure compost. Rose and Mercer reported a 100% degradation of parathion insecticide in cannery wastes. Singh reported a degradation efficiency of 96.03% for the endoslufin pesticide after 4 weeks, using composted soil with a moisture level of 38%. Al hattab and Ghaly achieved a captan removal efficiency of 92.4% in the first four days using hay compost.

Several researchers stated that polyhalogenated hydrocarbons, used in pesticides, can be metabolized under anaerobic conditions. However, other researchers noted that pesticides may largely persist unchanged during the composting process. Muller and Korte noted little to no degradation of the aldrin, dieldrin and monolinuron during the composting of sewage waste sludge. Strom noted the presence of chlordanein finished compost from various US municipalities.

Phytoremediation

In this method plants are used to contain and remove harmfull environmental contaminants as shown in Figure. Kruger et al. reported a degradation efficiency in atrazine of 65% after 9 weeks in soil where Kochi sp. was planted. Coats and Anderson reported that degradation of atrazine, metrolachlor and triflualin was enhanced in soils where the Kochi sp. plant grows. Olette et al. reported that the aquatic plants L. minor, C. aquatic and E. Canadensis were capable of removing 2.5% - 50% of dimethomorph and flazasulfuron present in the water. Buyanovsky et al. noted that the fungi rhizosphere was capable of degrading carbofuran by using it as its carbon source. Gordon et al. noted that 95% of trichloroethylene was removed from wastewater by hybrid polar trees during growing season. Stearman et al. noted that in constructed wetlands, cells with plants were capable of removing 77.1% and 82.4% of simazine and metolachlor, respectively, while cells without plants were only capable of removing 64.3% and 63.2%, respectively. Wang et al. reported that in the first 20 days of plant growth, oilseed rape seedlings were capable of removing 20% of chlorpyrifos pesticide.

Bioaugmentation

This method uses isolated microbes for the degradation of pesticides. The pesticides are quickly metabolized and converted to products with a lower toxicity under aerobic and anaerobic conditions. Some pesticides are capable of being degraded by certain bacterial strains as their sole nutrient or carbon source. Such microbial reactions are known as mineralizing because of the large amount of carbon dioxide released during metabolism.

Dichlorodiphenyltrichloroethane, normally resistant to biodegradation, has been reported to have been metaboilized in an anaerobic microbial culture, followed by an aerobic one. Bhadhade et al. reported that bacteria isolated form soil was capable of degrading 83% - 93% of the organophosphorous pesticide, monocrotophos. Ohshiro et al. reported a 96% reduction in isoxathion from the organophosphourus pestiside by bacteria isolated from turf green soil. Tang and You, reported that the triazophos bacteria is capable of degrading 33.1% to 95.8% of pesticides.

The down side of such a method is the establishment of the microbes in the presence of other microbial populations present in the contaminated soil. Acea et al. noted that the population of the introduced bacteria may be reduced due to susceptibility to predation or starvation.

Permissions

Index

www.ingramcontent.com/pod-product-compliance
Lightning Source LLC
Chambersburg PA
CBHW082037190326
41458CB00010B/3387